ELECTRICAL ENGINEERING DEVELOPMENTS

CAPACITORS

THEORY OF OPERATION, BEHAVIOR AND SAFETY REGULATIONS

ELECTRICAL ENGINEERING DEVELOPMENTS

Additional books in this series can be found on Nova's website under the Series tab.

Additional E-books in this series can be found on Nova's website under the E-book tab.

ENERGY SCIENCE, ENGINEERING AND TECHNOLOGY

Additional books in this series can be found on Nova's website under the Series tab.

Additional E-books in this series can be found on Nova's website under the E-book tab.

ELECTRICAL ENGINEERING DEVELOPMENTS

CAPACITORS

THEORY OF OPERATION, BEHAVIOR AND SAFETY REGULATIONS

KRISTOFER N. MULLER
EDITOR

NOVINKA

New York

For permission to use material from this book please contact us:
Telephone 631-231-7269; Fax 631-231-8175
Web Site: http://www.novapublishers.com

Library of Congress Cataloging-in-Publication Data

ISBN: 978-1-62417-585-5

Published by Nova Science Publishers, Inc. † New York

CONTENTS

PREFACE

In this book, the authors discuss the theory of operation, behavior and safety regulations of capacitors. Topics include (CCTO) ceramics as potential candidates for capacitor applications; devices, materials and mechanisms of capacitive sensors for gas and humidity detections; an overview of the leakage current mechanisms of ultra-thin gate dielectrics used in new-age capacitors and transistors; integration of capacitors into integrated circuits; and the impact of capacitor energization over the power quality of the MV distribution grid.

Chapter 1 - $CaCu_3Ti_4O_{12}$ (CCTO) ceramics are potential candidates for capacitor applications due to their large dielectric permittivity (ε') values of up to 300 000. The underlying mechanism for the high ε' is an internal barrier layer capacitor (IBLC) structure of insulating grain boundaries (GB) and conducting grain interiors (bulk). This behaviour is reviewed and discussed in detail. The origin of the IBLC structure is attributed to a small Cu non-stoichiometry in nominally insulating $CaCu_3Ti_4O_{12}$, which varies between the GBs and bulk. Such non-stoichiometry effects are studied in detail by analyzing bulk ceramics of different composition within the ternary CaO-CuO-TiO_2 phase diagram using X-ray diffraction (XRD), scanning electron microscopy (SEM) and impedance spectroscopy (IS). At least two defect mechanisms are suggested to exist. It is further shown that the development of the defect mechanisms in CCTO and the concomitant formation of the IBLC structure strongly depend on the processing conditions of CCTO ceramic pellets such as the sintering temperature. Nominally stoichiometric CCTO bulk ceramics sintered at different temperatures are analyzed using XRD, SEM and IS. The performance of CCTO ceramics for IBLC applications is controlled by subtle modifications in the compound stoichiometry that is strongly dependent on the ceramic sintering temperature.

Chapter 2 - Capacitive gas sensors are a distinct group of microelectronic devices that detect the analyte concentration by the change of capacitance. While currently resistor-based devices dominate the domain of solid-state gas detection, specific advantages allow gas sensing capacitors to be favored in many circumstances with promising prospects. Starting with an overview of gas detection, this chapter reviews the history and recent progresses of several kinds of capacitive gas or humidity sensors. Detailed discussions are carried out on the device compositions, sensing materials, and principles. Dielectric or electrode materials used and studied in different capacitive gas sensors including organic, inorganic, and novel nano-scale structures, as well as proved or proposed detection processes, especially the reversible interactions between the analyte and the sensing materials, are described along with the corresponding type of devices. Some non-capacitive solid-state or microelectronic sensors with identical or alike sensing mechanisms to the gas sensing capacitors will also be discussed. Besides, sensor design criteria and device fabrications are briefly introduced.

Chapter 3 - The lateral and vertical shrinkage of components in integrated circuits causes abrupt increase in the power consumption of the devices due to the tunneling of charge carriers across the dielectric layer. Novel materials to replace the conventional SiO_2 as the gate electrode have been actively researched in the last two decades, with the advantage of reducing the leakage currents by keeping the capacitive same as that of ultra-thin (<1 nm) SiO_2 films.

This chapter is an overview of the leakage current mechanisms of ultra-thin gate dielectrics used in new-age capacitors and transistors, and their reliability issues.

Chapter 4 - While Integrated circuits consume only 10% of the volume of a circuit-board, the majority of the space is occupied by passive elements such as capacitors, inductors or resistors. At system-level, integration of passive elements plays a crucial role in the miniaturization of functional devices such as mobile phones, computers and i-pads. However, integration of passive elements is technologically challenging, because some of these elements, for instance capacitors, can not compromise the reduction in the area of the device. Several smart technological innovations have been devised in the last decade to circumvent the "available reducing area in a chip", and this chapter gives an overview of the technology of integrated capacitors and their functioning.

Chapter 5 - The connection of capacitor banks at the MV distribution network busbar feeder is a technique widely used by the Spanish utilities. It decreases the reactive power transit (diminishing the losses) and improves the voltage profile along the distribution lines. However, this technique has a negative influence over the Quality of Supply because the capacitor's energization generates an oscillatory transient surge that propagates along the lines little damped which can cause problems in the MV elements and in the LV equipments. To analyse the generation and propagation of these surges, several power monitors were installed at different sites of the MV distribution network. A Fourier analyse of the waveforms recorded will show the frequencies at different sites revealing the 'natural' frequencies of the users (including the MV/LV transformer and their LV loads).

Finally, the techniques to reduce the amplitude and the negative effects of these surges will be shortly presented.

In: Capacitors
Editor: Kristofer N. Muller
ISBN: 978-1-62417-585-5

Chapter 1

$CaCu_3Ti_4O_{12}$ (CCTO) CERAMICS FOR CAPACITOR APPLICATIONS

Rainer Schmidt[*] and Derek C. Sinclair[#]

ABSTRACT

$CaCu_3Ti_4O_{12}$ (CCTO) ceramics are potential candidates for capacitor applications due to their large dielectric permittivity (ε') values of up to 300 000. The underlying mechanism for the high ε' is an internal barrier layer capacitor (IBLC) structure of insulating grain boundaries (GB) and conducting grain interiors (bulk). This behaviour is reviewed and discussed in detail. The origin of the IBLC structure is attributed to a small Cu non-stoichiometry in nominally insulating $CaCu_3Ti_4O_{12}$, which varies between the GBs and bulk. Such non-stoichiometry effects are studied in detail by analyzing bulk ceramics of different composition within the ternary CaO-CuO-TiO_2 phase diagram using X-ray diffraction (XRD), scanning electron microscopy (SEM) and impedance spectroscopy (IS). At least two defect mechanisms are suggested to exist. It is further shown that the development of the defect mechanisms in

---

[*] Rainer Schmidt: Universidad Complutense de Madrid, Departamento Física Aplicada III, GFMC, Facultad de Ciencias Físicas, 28040 Madrid, Spain. The University of Sheffield, Department of Materials Science and Engineering, Mappin Street, Sheffield S1 3JD, United Kingdom. Corresponding author. Electronic mail: rainerxschmidt@ googlemail.com

[#] Derek C. Sinclair: The University of Sheffield, Department of Materials Science and Engineering, Mappin Street, Sheffield S1 3JD, United Kingdom.

CCTO and the concomitant formation of the IBLC structure strongly depend on the processing conditions of CCTO ceramic pellets such as the sintering temperature. Nominally stoichiometric CCTO bulk ceramics sintered at different temperatures are analyzed using XRD, SEM and IS. The performance of CCTO ceramics for IBLC applications is controlled by subtle modifications in the compound stoichiometry that is strongly dependent on the ceramic sintering temperature.

I. Introduction

The ternary oxide compound $CaCu_3Ti_4O_{12}$ (CCTO) has been considered for applications as a capacitor material due to its exceptionally high dielectric permittivity values. In polycrystalline ceramics the highest values reported are ≈ 300 000, ≈ 700 for thin films and ≈ 100 000 for single crystals [1-12]. It is now relatively well established that such giant permittivity values are extrinsic in origin. In single crystals high permittivity has been interpreted as an extrinsic electrode – sample interface barrier [13, 14], whereas in polycrystalline CCTO ceramics it is commonly explained in terms of an internal barrier layer capacitor (IBLC) structure of thin insulating grain boundaries (GBs) and conducting grain interior (bulk) regions [15, 16]. Electrode–sample interface barriers can also contribute to the high permittivity in ceramics [17], which has been interpreted previously as a surface barrier layer structure (SBLC) [18]. Since GB regions are generally expected to cover the conducting grain interiors, IBLCs may in fact resemble a core-shell microstructure which is well known to exhibit large capacitance [19]. Conventional core-shell devices generally require sophisticated processing to fully cover conducting grains or an equivalent interior phase with an insulating secondary phase. Contrarily, the IBLC core-shell microstructure in CCTO may arise directly from subtle chemical differences between the CCTO phases being formed at GB and grain interior (bulk) regions during ceramic processing and a secondary non-CCTO GB phase is not present. However, the stoichiometric differences between GBs and bulk in CCTO are not yet fully understood despite considerable research efforts in recent years. The work presented here is therefore intended to shed light on the defect chemistry and non-stoichiometry of CCTO.

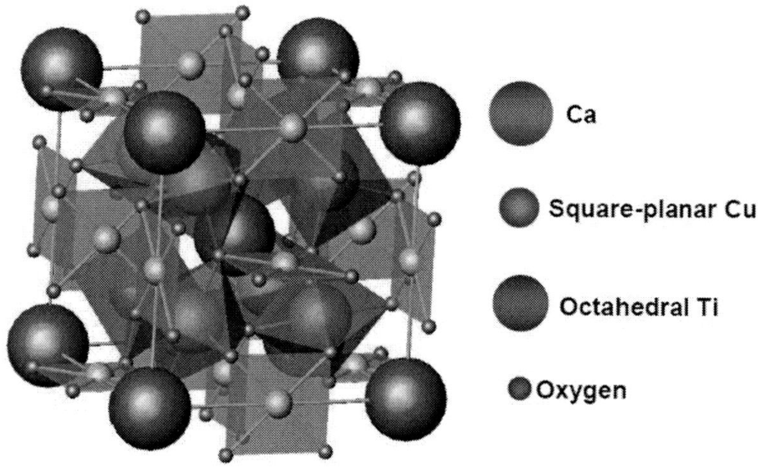

Figure 1. $CaCu_3Ti_4O_{12}$: A strongly distorted, A-site ordered double-perovskite structure.

In particular, the influence of processing conditions on the defect mechanisms and the formation of the IBLC structure is reported. This work comprises two experimental parts, where (A) ceramics with various compositions in the CaO-CuO-TiO_2 ternary phase diagram and (B) nominal CCTO ceramics sintered at various temperatures have been comprehensively investigated and characterized by a variety of techniques.

The crystal structure of CCTO is a 1:3 A-site ordered perovskite ($A'A''_3B_4O_{12}$). The oxygen octahedra are heavily tilted and the A" site Cu cations adopt a four-fold square-planar coordination. The unit cell is a doubled simple perovskite cell (Figure 1) and is commonly indexed as cubic Im-3. The octahedral tilting occurs because of the large cation size mismatch between the A'-site Ca^{2+} (1.34 Å) and the A"-site Cu^{2+} (0.57 Å) cations [20].

II. Impedance Spectroscopy (IS) and Non-Stoichiometry Models in CCTO

A. Basic Principles of Impedance Spectroscopy

Temperature dependent impedance spectroscopy (IS) enables different contributions to the dielectric and resistive properties of condensed matter to be deconvoluted and characterized separately [21, 22]. IS is therefore the method of choice to separately determine and investigate the dielectric properties of GB and grain interior (bulk) regions in electrically inhomogeneous electroceramics such as CCTO [4].

IS experiments consist of an electric stimulus in terms of a time (t)-dependent alternating voltage signal U of angular frequency ω and amplitude U_0 applied to a sample, and effectively the amplitude I_0 and phase shift δ of the current response signal I are measured.

$$U(\omega, t) = \boldsymbol{U_0} \cos(\boldsymbol{\omega} t); \qquad \rightarrow \qquad I(\omega, t) = I_0 \cos(\boldsymbol{\omega} t - \delta); \quad (1)$$

All parameters defined by the applied voltage signal (U_0, ω) are printed in blue/bold, the measured parameters of the current response are in red (I_0, δ). One phase of the applied voltage signal corresponds to a 2π (360°) rotation of the voltage arrow $U(\omega, t)$ shown on the phasor diagram in Figure 2.

The current response of ideal circuit elements is: (1) in-phase with the applied voltage in the case of an ideal resistor R; (2) out-of-phase by $\delta = -\pi/2$ for an ideal capacitor C; and (3) out-of-phase by $\delta = +\pi/2$ for an ideal inductor. All phase angles are time independent and all arrows rotate at constant angles for a given frequency.

The impedance can therefore be defined as a time-independent complex number Z^* (= Z' + iZ''), where the impedance fraction in-phase with $U(\omega,t)$ is defined as the real part, and the fraction +/- π/2 out-of-phase as the positive/negative imaginary part. The complex impedance Z^* definitions for several electronic circuit elements are given in Figure 2.

It is well established that different dielectric relaxation processes detected by IS such as those found at the GB and bulk regions can, in the simplest case, be understood in terms of the brick work layer model where each GB or bulk type contribution is described by one RC circuit element consisting of a resistor and capacitor in parallel [23].

Several dielectric relaxations in series can then be described by several RC elements connected in series, where the macroscopic impedance is just the sum of all series RC impedances.

The RC model works particularly well for insulators such as dielectrics, where the capacitor describes the ability of the material to store charge and the parallel resistor describes the leakage current due to some un-trapped charge carriers bypassing the ideal charge storage element.

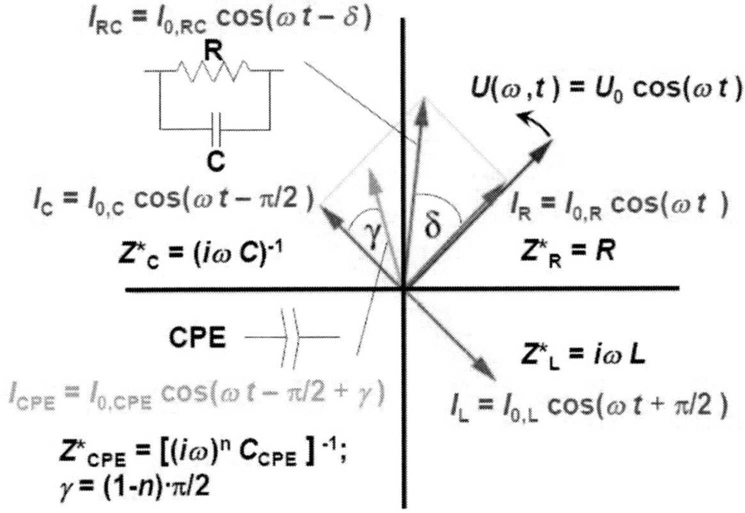

Figure 2. Impedance response of circuit elements on a phasor diagram: applied voltage stimulus $U(\omega t)$, current response I_R from an ideal resistor, I_C from an ideal capacitor, I_{RC} from an ideal parallel resistor-capacitor (RC) element with phase shift δ, I_L from an ideal inductor, and I_{CPE} from a constant phase element (CPE) describing a non-ideal capacitor with a frequency independent phase shift γ with respect to I_C.

In reality, the dielectric relaxation processes in electroceramics like CCTO are usually not ideal and, therefore, experimental IS data can usually not be modeled by ideal RC elements. To fit the data to an adequate equivalent circuit model the non-ideality in one specific dielectric relaxation can be accounted for in the respective RC element by connecting a constant-phase element (CPE) in parallel (R-CPE-C) or by replacing the ideal capacitor with a CPE (R-CPE). The phase angle and the complex impedance $Z_{CPE}*$ of a CPE are shown in Figure 2, where C_{CPE} is the CPE specific capacitance. The CPE capacitance obtained from fits to an R-CPE circuit can be converted into a real capacitance given in [Farad] using a standard procedure [24].

The critical exponent n defined in Figure 2 has typical values of $n = 0.6 - 1$. $n = 1$ constitutes the ideal case of an ideal capacitor for an ideal dielectric relaxation. In a non-ideal R-CPE-C or R-CPE circuit decreasing n values indicate a broadening of the respective dielectric relaxation peak as a reflection of the broadening of the distribution of relaxation times, τ, across the sample. In an ideal RC element τ is given by $\tau = RC$. The exact shape of the distribution of τ is complicated to be determined from IS data, and the exponent n constitutes a semi-empirical parameter to reflect an increasing width of the distribution in τ by decreasing n values. In addition to the complex impedance Z'-Z'', IS data can also be represented in alternative formats using the standard conversions shown in Figure 3 for the real part of the dielectric permittivity (ε') or capacitance (C') and the imaginary part of the dielectric modulus (M''). Such plots often contain important information on the resistance or permittivity of a specific relaxation. In a scenario of a ceramic sample such as CCTO with one GB and one bulk (b) relaxation as represented by two RC elements connected in series the following spectral features are expected as shown from the simulations presented in Figure 3: (1.) In plots of Z'' vs Z' two semicircles appear with each semicircle diameter corresponding to the respective resistance R, (2.) Z'' vs frequency (f) plots show two relaxation peaks with peak heights of $R/2$ for each relaxation, (3.) C' vs f or ε' vs f display two approximately f independent C' (ε') plateaus C_{GB}* and C_b*, that correspond approximately to the GB (C_{GB}) and bulk capacitance (C_b) and show an intermediate sharp drop at a distinct f, (4.) ε'' vs ε' plots display one semicircle with the diameter being directly proportional to the GB permittivity (not shown), and (5.) M'' vs f displays two relaxation peaks with the peak heights being inversely proportional to each capacitance, $1/C$.

The resistance and capacitance values used for the simulations are given in the Figure 3 caption, where the values were chosen to represent a realistic scenario of one extrinsic GB- and one intrinsic bulk-type relaxation.

In the framework of the brick work layer model of two RC elements connected in series, the GB and bulk capacitance plateaus C_{GB}* and C_b* depicted in the C' vs f plot in Figure 3 are both only an approximation for the GB and bulk capacitance C_{GB} and C_b respectively. In fact, both plateaus are composite terms, where C_{GB}* contains all resistor and capacitor terms and C_b* contains a contribution from the GB capacitance. This is apparent from Figure 3, since C_{GB}* and C_b* do not coincide exactly with the values of the capacitors C_{GB} (10 pF) and C_b (1 pF). The exact expressions for C_{GB}* and C_b* are given by equations (2) and (3) respectively.

$$C^*_{GB} = \frac{R^2_{GB} C_{GB} + R^2_b C_b}{(R_{GB} + R_b)^2} \qquad (2)$$

$$C^*_b = \frac{C_{GB} \times C_b}{C_{GB} + C_b}; \qquad (3)$$

In the dielectric permittivity format of ε_{GB}*, ε_b*, ε_{GB} and ε_b, equations (2) and (3) are equivalent. In the case that the GB and bulk resistance R_{GB} and R_b are sufficiently different, i.e. $R_{GB} \gg R_b$, then C_{GB}* constitutes an excellent estimate for C_{GB}. This is usually the case for CCTO where R_{GB} and R_b commonly vary by more than 3 orders of magnitude. C_b* is a good estimate for C_b for the case where the two capacitors C_{GB} and C_b are sufficiently different, i.e. $C_{GB} \gg C_b$. This is not always the case for CCTO as is discussed below in section IV./A.

In CCTO with distinctively different R_{GB} and R_b values the *M*" vs *f* plots are particularly useful to display the bulk relaxation peak. *Z*" vs *f* plots highlight the relaxation peak with the highest resistance *R* (e.g. the GB peak in CCTO), whereas the *M*" vs *f* plots highlight the smallest capacitance (i.e. the bulk peak in CCTO).

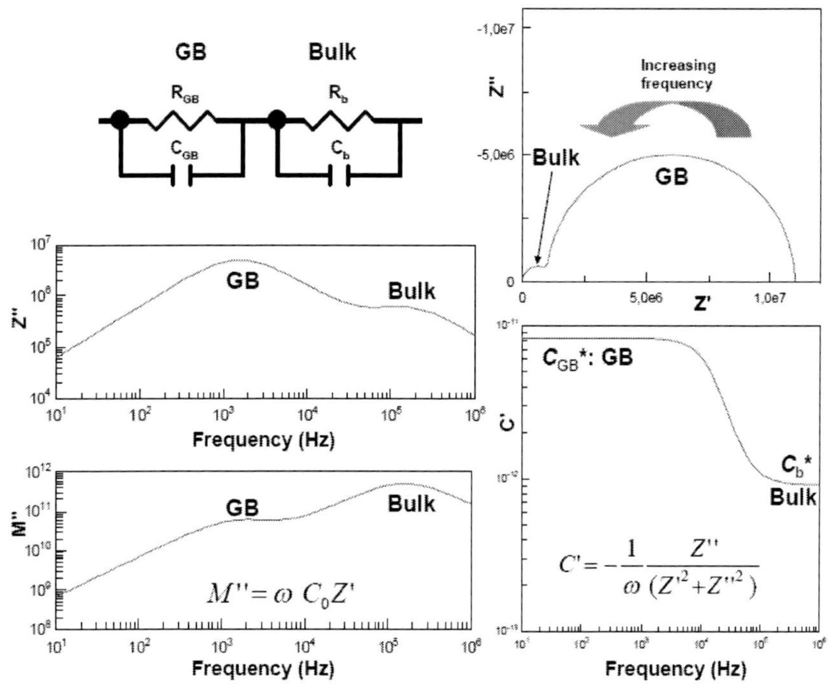

Figure 3. Simulated IS data for two RC elements connected in series, presented in different formats: *Z*" [Ω] vs frequency (*f*), *M*" vs *f*, *Z*" [Ω] vs *Z*' [Ω] and *C*' [Farad] vs *f*. Simulations were carried out with R_{GB} = 10 MΩ, R_b = 1 MΩ, C_{GB} = 10 pF, C_b = 1 pF.

The approximate peak frequencies for GB or bulk relaxation peaks as depicted in Figure 3 are given in equation 4 for the respective plots of -*Z*'' vs *f* (GB) and *M*'' vs *f* (bulk), using two ideal RC elements connected in series. Equation (4) implies an extrinsic GB-type relaxation with large resistance and capacitance to appear at lower *f* than the bulk, which is a common feature in experimental impedance spectra from electroceramics.

The different ordinates of the GB and bulk peaks are indicated in equation (5) for a CCTO scenario with distinct differences between the GB and bulk RC elements.

$$f_{\max}(-Z'') \approx \frac{1}{2\pi\ R_{GB} C_{GB}};\ f_{\max}(M'') \approx \frac{1}{2\pi\ R_b C_b} \tag{4}$$

$$-Z''(f_{\max}) \approx \frac{R_{GB}}{2};\ M''(f_{\max}) \approx \frac{C_0}{2C_b} \tag{5}$$

C_0 represents the capacitance of the empty measuring cell in vacuum. The differences in the peak ordinates in equation (5) guarantee that in -*Z*'' vs *f* the relaxation peaks representing the largest resistance (GB) and in *M*'' vs *f* the peaks representing the smallest capacitance (bulk) are the most strongly pronounced (see Figure 3).

In the case where both resistors in Figure 3 are thermally activated and follow Arrhenius type behavior, as is the case for CCTO, the dielectric relaxation peaks and the drop in *C*' depicted in Figure 3 would all move to higher *f* with increasing temperature, *T*. This appears as if the full spectrum shifts to higher *f* as is demonstrated below in Figure 4b for the example of a *C*' vs *f* plot.

B. Impedance Spectroscopy on CCTO

The typical ceramic IBLC structure for CCTO ceramics is depicted in Figure 4a, where the semiconducting grain interiors (bulk) are shown in dark/blue whereas the surrounding insulating GB regions are colourless. No percolation path for high conductivity is available and the macroscopic direct current (dc) resistance is insulating. By carrying out IS it appears that at sufficiently high *f* a sharp drop occurs in *C'* vs *f* plots and the giant extrinsic GB permittivity reduces to the bulk relative dielectric permittivity of ≈ 100 (Figure 4b) [18, 25]. This is the typical behaviour of an IBLC structure, where the sharp *C'* drop occurs when the mean electron conduction path permitted by the applied alternating voltage signal decreases below the average grain size at increased *f*. The low-*f* dielectric response is therefore dominated by the insulating GB and the high-*f* response by the conducting bulk contribution. Both dielectric relaxation processes (GB and bulk) are represented in an idealised way by two RC elements connected in series (Figure 4c). In CCTO ceramics the inhomogeneous dielectric behaviour can be pronounced: GB and bulk resistance can differ by a factor of up to $\approx 10^5$. This is demonstrated in Figure 5, where the GB and bulk (b) semicircles are not both visible at the same temperature. Only at high *T* (300 K) does the GB exhibit sufficiently low resistance to be fully developed as an arc in the *Z''* vs *Z'* plot, whereas the bulk resistance is small and the bulk relaxation peak occurs at a higher *f* than the measured spectrum.

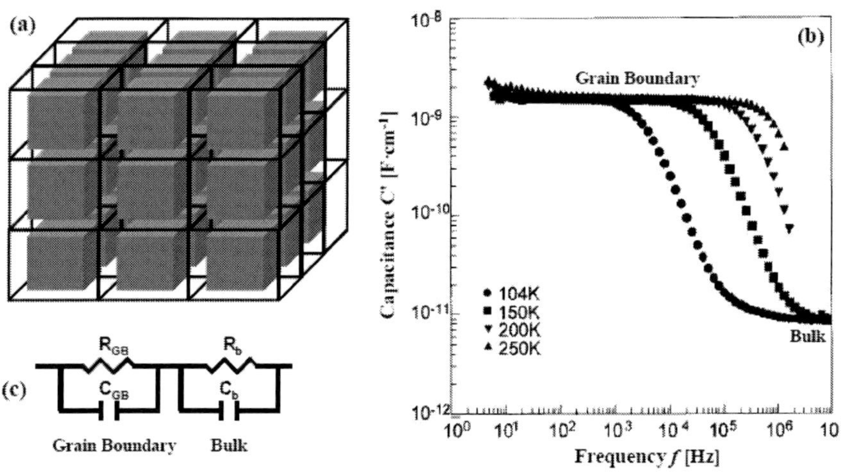

Figure 4. Internal barrier layer capacitor (IBLC) structure for CCTO: (a) Ceramic grains are represented schematically by cubes (solid lines). The semiconducting (grain) bulk regions are represented by smaller (blue) cubes. The surrounding insulating grain boundary areas are colourless. (b) Impedance spectroscopy (IS) data from CCTO ceramics represented in the notation of the real part of capacitance *C'* vs *f*. Image reproduced from [15], with permission from the American Institute of Physics (AIP). (c) Idealized equivalent circuit model to account for GB and bulk (b) dielectric relaxation processes.

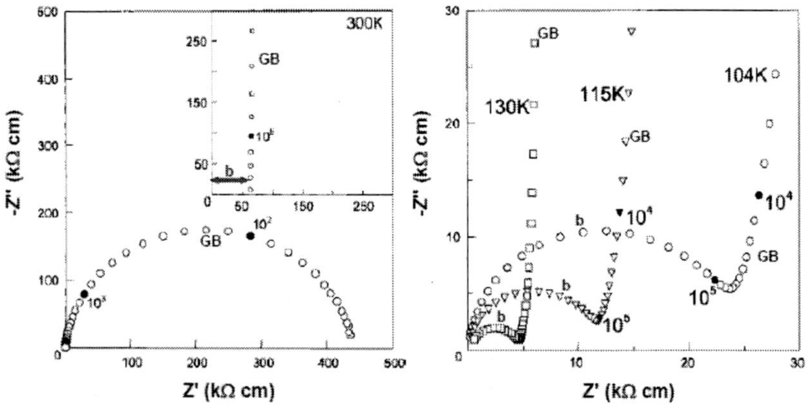

a) b)

Figure 5. Complex impedance Z'' vs Z' plots for ceramic CCTO. GB and bulk (b) contributions are indicated. (a) At 300 K only the GB semicircle is visible, where the bulk (b) only appears as a non-zero intercept of the data curve with the Z' axis at high f (see bulk arrow (b) in the Figure inset). (b) On cooling the bulk relaxation peak shifts to lower f and is visible at 104, 115 and 130 K. The change in semicircle diameter indicates thermally activated charge transport. The GB semicircle is only partially visible due to high R and cannot be resolved. Images reproduced from [15] with permission from the American Institute of Physics (AIP).

Only a non-zero intercept of the data with the Z' axis appears at the upper limit of f as demonstrated by the (red) bulk (b) arrow in the Figure 5 inset.

At low T (104, 115 and 130 K) the bulk resistance is larger, the bulk relaxation peak has shifted to lower f in agreement with equation (4) and can now be clearly resolved. On the other hand, the GB resistance is now too large to be resolved in the form of a fully developed GB semicircle and only the onset is visible. The decreasing size of the bulk semicircle in Figure 5(b) with increasing T is indicative of thermally activated charge transport. The associated charge transport activation energies, E_a, associated with either GB or bulk CCTO conduction processes, have been reported to differ up to a factor of $\approx$ 10 [16, 26]. This is highlighted in Figure 6, where R_{GB} and R_b values were determined for a CCTO ceramic sample from the respective semicircle diameters in Z'' vs Z' plots.

The data are plotted in the conductivity format of $1/R_{GB}$ and $1/R_b$. The large difference displayed in E_a as well as in the nominal resistance between R_{GB} and R_b is a strong indicator that different charge transport mechanisms must exist. This in turn implies that also perceptible chemical differences between GB and bulk regions must exist to cause this variation.

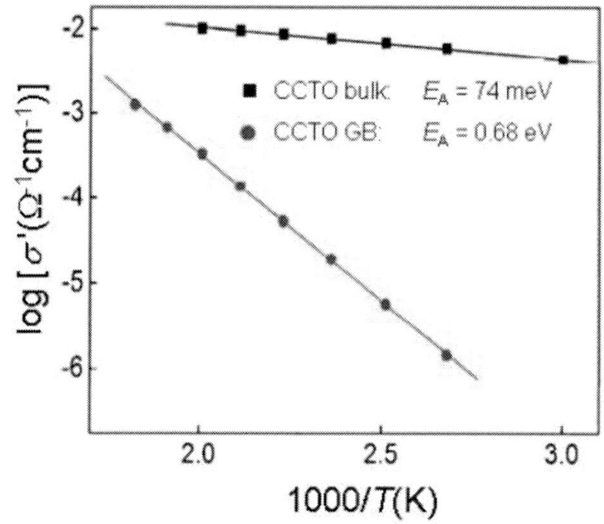

Figure 6. Arrhenius plots of GB and bulk conductivities for CCTO. The GB and bulk activation energies E_A as well as the nominal conductivity values differ significantly. Image reproduced from [27], with permission from M. Li.

These chemical differences between GB and bulk CCTO are not well understood, especially the low bulk resistance in nominally stoichiometric (and therefore insulating) $CaCu_3Ti_4O_{12}$. This has led to controversial discussions in the literature over potential defect mechanisms and the non-stoichiometry that are responsible for inducing semiconductivity in the grains.

C. Non-Stoichiometry Models in CCTO

In the literature several controversial reports exist, where different defect mechanisms and associated solid solutions have been proposed to explain CCTO bulk semi-conductivity and the IBLC structure, reporting on various cation distributions and different types of charge transport (n-type electron or p-type hole conduction).

Such conduction may rely on electron transport based on mixed valence Ti^{3+}/Ti^{4+} or Cu^{1+}/Cu^{2+} (n-type), or Cu^{2+}/Cu^{3+} (p-type). The following defect mechanisms, solid solutions and possible compensation mechanisms have been reported in the literature [28-30] or appear to be plausible from the results presented here; brackets ( ) indicate A'' Cu sites:

(A) $O_o^{2-} + 2Ti^{4+} \rightarrow V_0^{''} + 2Ti^{3+} + 1/2O_2 \uparrow$: $Ca(Cu_3)Ti_{4-u}^{4+}Ti_u^{3+}O_{12-u/2}$

(B) $3Cu^{2+} \rightarrow 2Cu^{1+} + Ti_{Cu}^{4+} + Cu \uparrow$: $Ca(Cu_{3-3v}^{2+}Cu_{2v}^{+}Ti_v^{4+})Ti_4O_{12}$

(C) $Cu^{+} + Ti^{4+} \rightarrow Cu^{2+} + Ti^{3+}$: $Ca(Cu_{3-v}^{2+}Ti_v^{4+})Ti_{4-2v}^{4+}Ti_{2v}^{3+}O_{12}$

(D) $3Cu^{2+} \rightarrow V_{Cu}^{''} + 2Cu^{3+} + Cu \uparrow$: $Ca(Cu_{3-3y}^{2+}Cu_{2y}^{3+}V_y^{''})Ti_4O_{12}$

(E) $O_o^{2-} + Ti^{4+} \rightarrow V_0^{''} + Cu_{Ti}^{2+} + 1/2O_2 \uparrow$: $Ca(Cu_3^{2+})Ti_{4-w}^{4+}Cu_w^{2+}O_{12-w}$

(A) represents oxygen loss compensated by Ti^{4+} reduction. (B) represents Cu loss or deficiency compensated by partial reduction of Cu^{2+} to Cu^{1+} and partial occupation of Cu sites by Ti^{4+}. Processes (A) and (B) would most likely take place during heat treatment at elevated temperatures. In (C) the Cu^{2+} reduction at elevated temperatures described in (B) is now assumed to be fully reversible upon cooling by reoxidation, whereas the Cu deficiency persists. The reoxidation of Cu^{1+} to Cu^{2+} in turn is compensated by partial reduction of Ti^{4+} to Ti^{3+}; i.e. an internal redox process takes place. Only the redox process is displayed in (C), the high temperature Cu reduction is identical to (B). (D) represents Cu loss due to Cu segregation or volatilisation and compensation is via partial oxidation of Cu^{2+} to Cu^{3+}. (E) corresponds to Cu excess and compensation is by Cu^{2+} occupying Ti sites and the formation of oxygen vacancies. Alternatively, compensation in (E) could also occur via Cu^{2+} oxidation to Cu^{3+} and fewer oxygen vacancies would form.

Models (A), (B) and (C) are supported by findings of n-type conduction [3]. Process (B) is consistent with reports of Cu^{1+} from XPS studies [31]. (D) is supported by findings of (I.) Cu loss [30, 32]; (II.) the presence of Cu^{3+} from XPS studies [30], and (III.) p-type bulk semiconductivity determined from thermopower measurements [8]. (E) has not been mentioned in the literature but may be applicable as discussed below.

D. Doped CCTO

One or more of the abovementioned defect mechanisms (A - E) must be present in CCTO ceramics with an IBLC structure to explain chemical differences between insulating GBs and semi-conducting bulk. In the literature a large amount of publications can be found to demonstrate that chemical doping can induce new or possibly modify one or more of the abovementioned defect mechanisms leading to distinctively varied IBLC structures with modified resistance and permittivity values of GB and grain interior regions. In Figure 7 two examples are given: A'-site Sr-doping in $Ca_{1-x}Sr_xCu_3Ti_4O_{12}$ [5] and A"-site Mn doping in $CaCu_{3-x}Mn_xTi_4O_{12}$ [28]. A'-site Ca substitution by Sr leads to a modest increase in GB and bulk capacitance, whereas A"-site Cu substitution by Mn has a much more dramatic effect: on the right hand panel of Figure 7 the dielectric inhomogeneity of the ceramic is shown to disappear, whereby the bulk resistance increases [33]. Data is plotted as *C'* vs *T*, where a step-like increase is displayed from the lower bulk (b) to the higher GB plateau for un-doped CCTO ceramics. Such an increase is missing in $CaCu_{2.94}Mn_{0.06}Ti_4O_{12}$; the ceramic is homogeneous, only the bulk capacitance is visible, the grains are highly resistive [27] and the defect mechanism leading to semiconductivity in undoped CCTO is suppressed or absent. This may lead to significant conclusions:

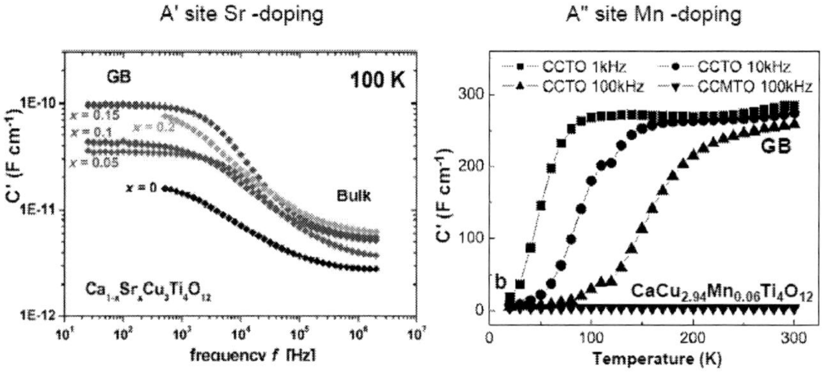

Figure 7. Effects of A'-site Sr- and A"-site Mn-doping on the dielectric properties of CCTO. GB and bulk (b) capacitance plateaus are indicated. Left panel: *C'* vs *f* plots for $Ca_{1-x}Sr_xCu_3Ti_4O_{12}$ ceramics with various *x* doping level. Right panel: *C'* vs *T* plots for $CaCu_{2.94}Mn_{0.06}Ti_4O_{12}$. Images reproduced from [5] with permission from the American Chemical Society (ACS) and from [28] with permission from the American Institute of Physics (AIP).

If the CCTO bulk semi-conductivity can be "killed" by Mn-doping, this is a clear indication that the differences between GB and bulk regions in CCTO may well emerge from subtle chemical differences in a GB-type and a bulk-type CCTO phase and not from a secondary non-CCTO phase. Such a secondary non-CCTO phase needed to be a GB-phase with a small volume fraction, because the CCTO bulk is commonly XRD phase-pure. The non-CCTO GB phase would be expected to fully cover the semi-conducting grain interior regions and not permit a conducting path across a ceramic sample. Such a phase has never been observed unambiguously and is in fact not plausible regarding the effects of Mn-doping. From Figure 7 it should be noted that the step-like decrease in C' vs f (left panel) and the step-like increase in C' vs T (right panel) may be somewhat opposing trends, but they are both fully consistent with a circuit dominated by two RC elements connected in series for GBs and bulk, as was pointed out in a recent simulation study [34].

In addition to the two examples shown in Figure 7, further articles in the literature report on an increase in the intrinsic bulk dielectric permittivity in $Na_{1/2}Bi_{1/2}Cu_3Ti_4O_{12}$ [35], an increase in GB permittivity in $CaCu_3Ti_{4-x}Nb_xO_{12+x/2}$ [36], and a decrease in GB permittivity in $La_{2/3}Cu_3Ti_4O_{12}$, $Y_{2/3}Cu_3Ti_4O_{12}$ and $Bi_{2/3}Cu_3Ti_4O_{12}$ [37].

In a recent and very comprehensive study it has been shown that the full rare-earth (RE) series exists [$(RE)_{2/3}Cu_3Ti_4O_{12}$], where the ceramics of this series all show an IBLC structure with the Pr containing ceramic showing the highest GB capacitance [38]. There are further examples in the literature reporting on a wealth of different doping mechanisms such as A'-site La-doping [39], A"-site Mg-doping [40], and B-site Ta, Zr, Cr, Al- or Mn-doping [41-46]. The summary in Figure 8 for several doping cations shows the distinct influence of the dopant on the IBLC structure in terms of variations in R_{GB} and R_b [26].

III. DEFECT CHEMISTRY WITHIN THE CaO-CuO-TiO_2 PHASE TRIANGLE

A. The CaO-CuO-TiO_2 Ternary Phase Diagram

Powders of various compositions within the ternary CaO-CuO-TiO_2 phase diagram (Figure 9) were synthesised by repeated grinding and heating at 1000 °C from different amounts of dried precursor oxides of $CaCO_3$, CuO and TiO_2.

All such compositions are indicated by (black) dots in Figure 9. A small portion of each powder was heated at 1100 °C for 12 h to enable investigation of the phase diagram for materials fired at both 1000 and 1100 °C. The latter is the most frequently used ceramic densification-sintering temperature, whereas CCTO powder synthesis is most commonly achieved at ≈ 1000 °C.

All labelled (numbered) compositions in Figure 9 were investigated in more detail by XRD lattice parameter determination and IS was carried out on pressed pellets sintered at 1100 °C. For these compositions (1 – 8) powder synthesis and pellet sintering were carried out simultaneously to exclude differences in the processing conditions.

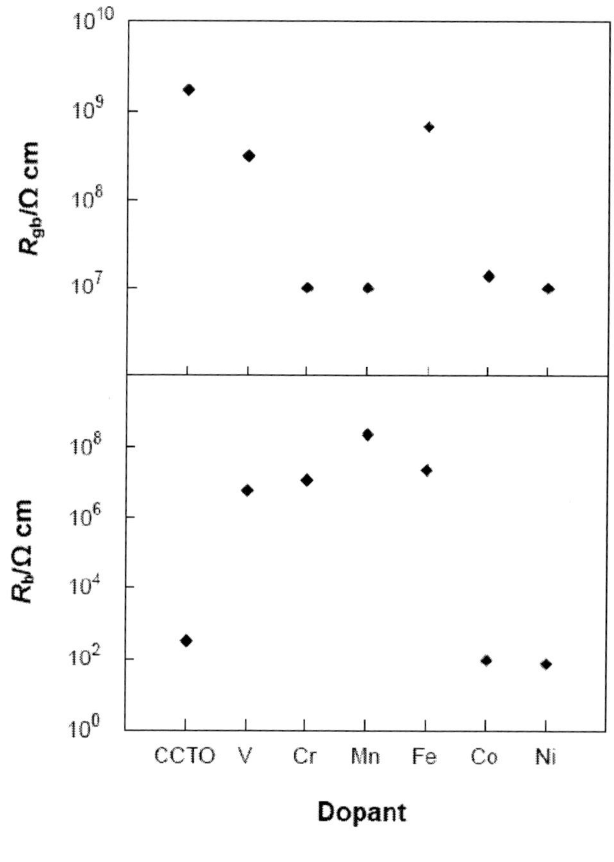

Figure 8. Comparison of R_{GB} and R_b at room temperature for undoped CCTO and x = 0.08 doped compositions, with the doping element shown. Images reproduced from [47], with permission from M.C. Ferrarelli.

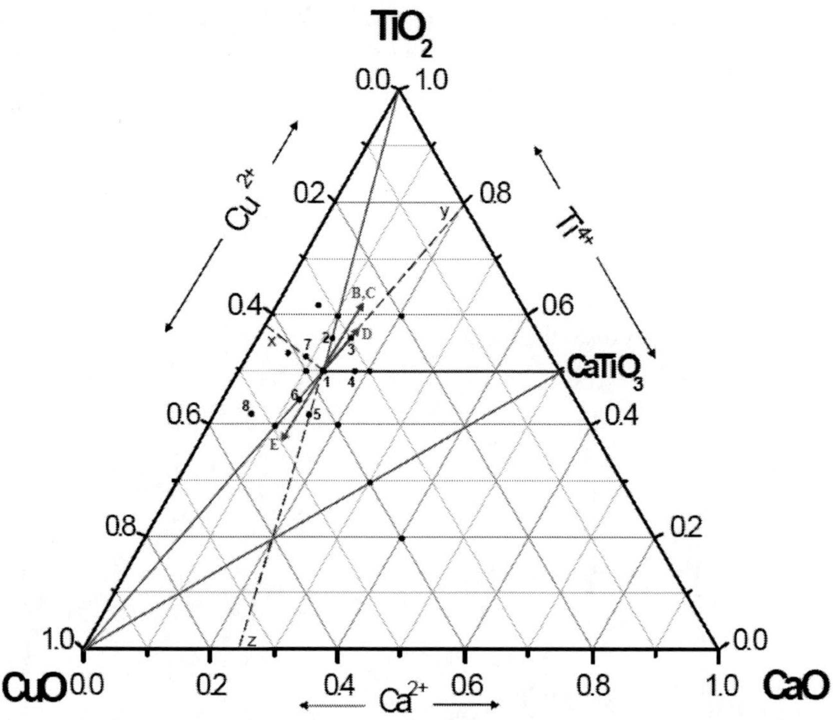

Figure 9. Ternary CaO-CuO-TiO_2 phase diagram. Black dots represent the compositions investigated. Numbers represent compositions analyzed in more detail by XRD and IS. Dashed blue lines represent CCTO compositions with potential Ca, Cu or Ti loss: $Ca_{1-x}Cu_3Ti_4O_{12}$, $CaCu_{3-y}Ti_4O_{12}$, $CaCu_3Ti_{4-z}O_{12}$, respectively. Green arrows indicate the directions of the potential solid solutions mentioned above in section II./C.

Compositions 1 - 8 in Figure 9 were chosen in a way such that the volume concentration of potential secondary phases like CuO, TiO_2 and $CaTiO_3$ would always be well below the percolation threshold for secondary phases randomly distributed within the primary CCTO matrix. This would ensure the dielectric response in all labeled compositions to always be dominated by the CCTO phase. The theoretical percolation threshold was calculated to be about 16% [48], whereas experimental values are commonly reported to be up to ≈ 20% [49].

The only single-phase composition detected within the main triangle was the CCTO phase, marked as composition 1. $CaTiO_3$ (CTO) was the only significant single-phase on the CaO-TiO_2 binary joint. Along all the solid (red) lines within the main triangle two phases are present, which are the two located on both ends of each line: e.g. CCTO and TiO_2 are present on the line connecting CCTO (composition 1) and the TiO_2 corner. The various secondary triangular regions within the main triangle all contain three phases, which are the ones that span out the secondary triangles: e.g. CCTO, CTO and TiO_2 are present inside the triangle with corners CCTO, CTO and TiO_2. No direct evidence for two CCTO compositions was found, i.e. for GB and bulk CCTO phases, in agreement with recent work [50]. The phase relationships detected are consistent with the notion that no obvious solid solution is evident and CCTO and CTO are the only non-trivial single-phase compositions. Potential solid solutions causing non-stoichiometry must therefore be small.

The XRD patterns for all numbered compositions in Figure 9, heat treated at 1100 °C, are displayed in Figure 10. The phase diagrams for heat treatments at 1000 and 1100 °C were found to be identical with one exception: no crystalline CuO phase could be detected at 1100 °C in any composition by powder XRD, whereas at 1000 °C crystalline CuO was detected in all compositions where it was expected according to the phase diagram. It is suggested that CuO may partially melt at 1100 °C, which was confirmed on CuO-rich pellets where a Cu-rich phase precipitated out of the pellets during sintering at 1100 °C and remained on the Pt foil after cooling. The absence of CuO phases in these pellets is clearly demonstrated in the XRD pattern presented in Figure 10.

In section II./.C. several solid solutions constituting potential CCTO defect mechanisms were mentioned to explain differences in GB and bulk regions, where all such mechanisms are indicated by (green) arrows in Figure 9. For demonstration purposes the length of the (green) arrows is depicted as being much higher than the expected small extent to which the solid solutions would be developed. Problems were encountered with the triangle having corners CuO, CTO and CaO, where the phase relationships could not be determined unambiguously. The compositions in this triangle were subjected to partial melting of CuO and possibly Ca containing phases other than CaO were formed. The single-phase compound Ca_2CuO on the CaO-CuO binary joint was postulated to exist in a recent work on powders heated at 950 °C [50] but could not be confirmed here for powders prepared at 1000 and 1100 °C.

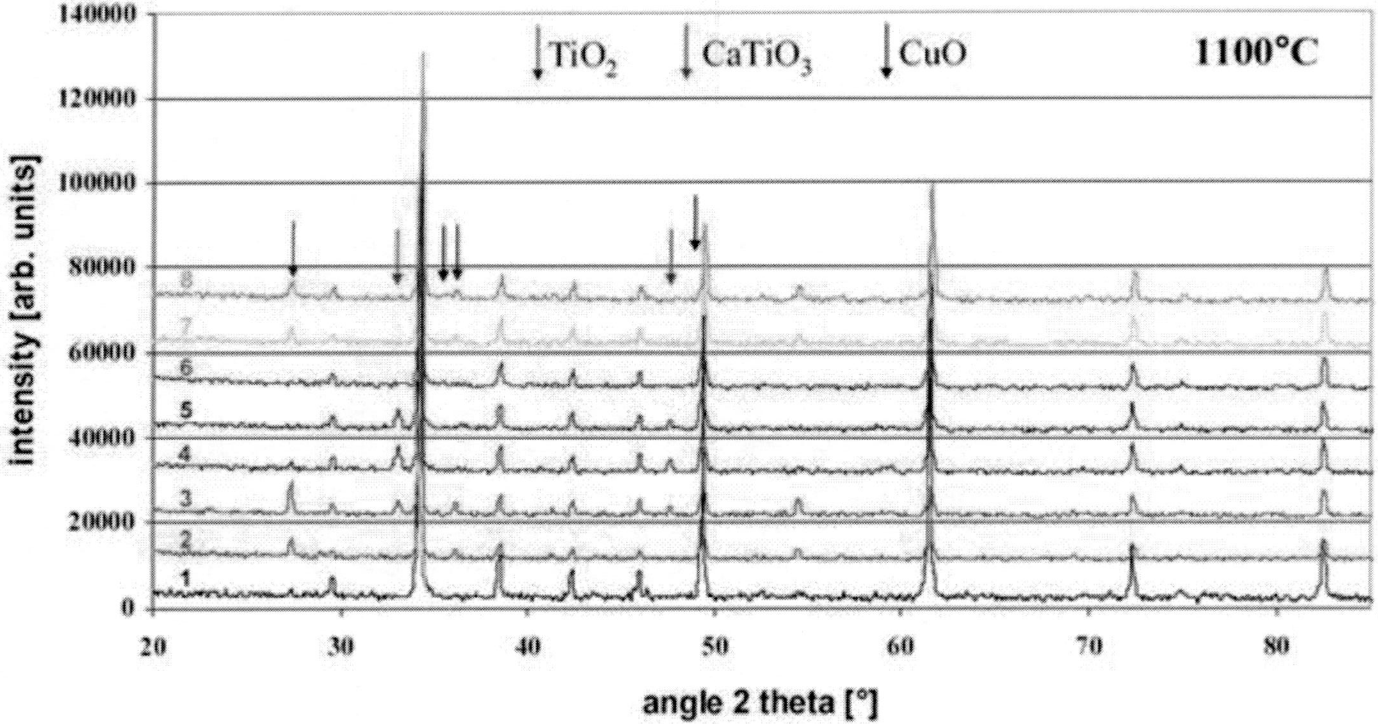

Figure 10. Powder XRD pattern of compositions 1-8 fired at 1100°C. Blue, red and black arrows correspond to the most intense TiO_2, $CaTiO_3$ and CuO reflections respectively.

B. Impedance Spectroscopy on Various CaO-CuO-TiO_2 Compositions

Pellets pressed from powders of compositions 1 – 8 were sintered at 1100 °C, polished and covered on both sides with Au electrodes using dc sputtering for IS measurements between 10 - 500 K at 1 Hz – 2 MHz. Figure 11 shows the data collected at 80 K in the format of the real part of dielectric permittivity ε' vs frequency f (Figure 11a) and imaginary part of the dielectric modulus M'' vs f (Figure 11b).

The impedance data for compositions 1 - 5 were fitted to the equivalent circuit model shown in Figure 11a, which had been successfully applied before for CCTO ceramics [18]. The intrinsic bulk dielectric relaxation process was fitted with a non-ideal R-CPE-C circuit (R1-CPE1-C1). The use of this circuit has been shown previously to be appropriate to account for non-ideal bulk relaxations in electroceramics [51]. The GB relaxation was fitted with a well-established non-ideal R-CPE element (R2-CPE2) and the data and model in Figure 11aandb show good agreement. A successful equivalent circuit fit was also obtained for composition 6; however, the fitting errors were larger, which may be due to less ideal dielectric behaviour.

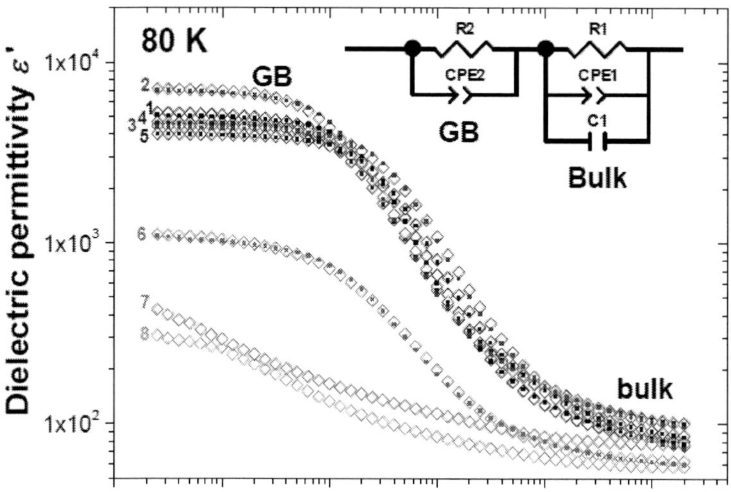

a)

Figure 11. (Continued)

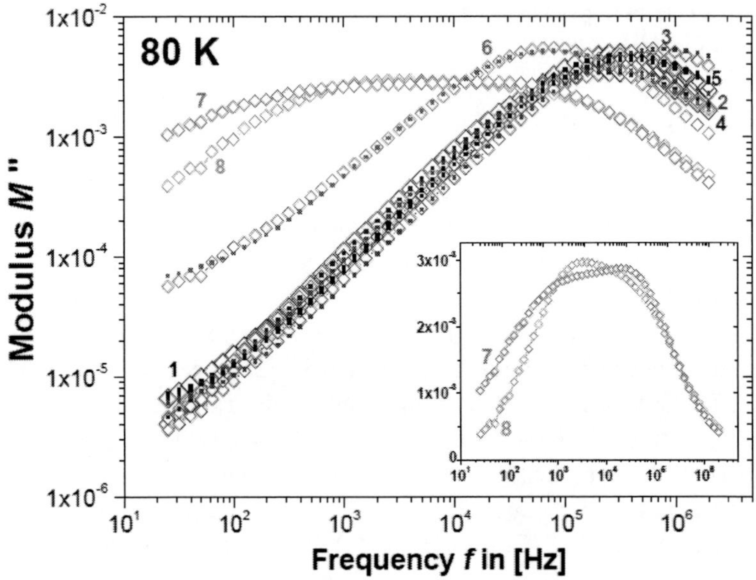

b)

Figure 11. IS data for compositions 1 - 8 at 80 K (a) ε' vs f, and (b) M'' vs f. Open diamonds (◊) are experimental data, full dots (▪) are fits to the circuit model shown in the inset of Figure 11a.

Figure 11a displays low and high-f plateaus as a result of electronic heterogeneity, which had been mentioned above to represent GB (ε_{GB}) and bulk (ε_b) permittivity respectively. It is obvious that different secondary phases have a distinct influence on the dielectric behaviour of CCTO and the IS data fall into two categories: Compositions 7 and 8 vary from 1 – 5, whereas 6 shows intermediate dielectric response and may fall in between. According to the phase diagram in Figure 9, compositions 7 and 8 belong to the same compositional triangle (CuO-CCTO-TiO_2), which would be a plausible explanation for their dielectric distinction. Below in section III./C. it will be shown that this interpretation is incomplete: compositions 5 and 6 also vary from 1 - 4 if their dielectric response is being related to the CCTO lattice parameter ***a***. It appears that it is in fact the presence or absence of the CuO secondary phase that separates the dielectric response into the two categories: i.) 1, 2, 3, 4 (no extra CuO), and ii.) 5, 6, 7 ,8 (extra CuO). According to Figure 11a the IBLC structure seems to be developed in all compositions, but to different extents. Such variations will be analysed in detail in the following to obtain information on potential defect mechanisms being present.

In several previous studies the effect of secondary phases on the CCTO giant dielectric permittivity has already been analysed in detail, although the giant permittivity is extrinsic in nature. This analysis therefore gives phenomenological information only and the investigation of the intrinsic bulk dielectric relaxation is required to gain fundamental insight into the potential intrinsic changes of the CCTO phase induced by the presence of secondary phases. As mentioned above in section II./A. the CCTO intrinsic bulk dielectric relaxation processes may be investigated best by plots of M'' vs f, which are shown in Figure 11b for compositions 1 – 8. The equivalent two categories are observed as in Figure 11a. For compositions 1 – 5 one high f relaxation peak each is displayed, which is indicative of a conventional bulk dielectric relaxation process. Compositions 7 and 8 display a flat or in fact a double peak structure in M'' vs f (see Figure 11b inset) and the bulk dielectric response seems to be distinctively different, whereas composition 6 may fall in between. Again, this interpretation is incomplete as mentioned above and the categories are i.) 1 - 4 and ii.) 5 - 8.

The rather untypical and strongly overlapping "double-peaks" in 7 and 8 both occur at significantly lower f when compared to the bulk peaks for 1 – 5, whereas the peak heights (ordinates) are only slightly reduced. Using equation (4) this implies that the two double-peak relaxation processes in 7 and 8 both exhibit significantly increased resistance as compared to the single CCTO bulk relaxation for compositions 1 – 5. The relaxation peak of composition 6 indicates increased resistance as well. From these findings it seems plausible that Cu-rich compositions may display higher resistance and an additional CuO phase drives the bulk CCTO peak to higher f towards the GB relaxation. This may serve as an explanation for the formation of the IBLC structure: in a recent study it has been shown that Cu segregates out of the CCTO grains during the sintering process [52], which can lead to Cu-rich resistive GBs and Cu-deficient conducting bulk regions.

Since the typical high f bulk dielectric relaxation peak displayed in 1 – 5 is missing in 7 and 8 (Figure 11b), it may be argued that the peak at higher f of the two double-peaks is more likely to represent the intrinsic bulk CCTO relaxation of composition 7 and 8. The origin of the second peak at slightly lower f is unclear and may represent a modified GB contribution or a secondary phase. A complete absence of the CCTO bulk relaxation is unrealistic due to the XRD patterns clearly displaying the CCTO phase, which in fact strongly dominates over all secondary phases (see Figure 10). Due to the double- peak overlap and considerable peak broadening the data for 7 and 8 cannot be fitted by an equivalent circuit model based on RC elements as for 1 – 6.

C. Combined Analysis of Lattice and Dielectric Parameters

Summarising results from the above section, indications have been found in the impedance spectra that potential non-stoichiometry in CCTO may involve Cu. In this section clear trends of bulk resistance R_b vs lattice parameter $\boldsymbol{a}$ and bulk dielectric permittivity ε_b vs $\boldsymbol{a}$ are presented that reveal the existence of at least two different defect mechanisms and associated solid solution in CCTO.

1. Solid Solution in Cu-Deficient Compositions

Figure 12a shows a plot of bulk resistance R_b obtained from equivalent circuit fits vs the lattice parameter $\boldsymbol{a}$, for compositions 1 - 6. $\boldsymbol{a}$ was determined from high resolution powder XRD pattern, where all diffraction peak positions were normalized to a Si standard and at least 4 peaks at high diffraction angles were used in each case to obtain $\boldsymbol{a}$ from a fitting routine using commercial WinXPOW® software. Dielectric data were collected from pellets sintered at 1100 °C. For compositions 1 – 4, R_b at 150 K shows a modest and approximately linear increase with $\boldsymbol{a}$. It is unclear at this point whether this increase of R_b with $\boldsymbol{a}$ represents a solid solution because the increase is small and may simply be a consequence of the increasing atomic distance of the cations. With increasing $\boldsymbol{a}$ the orbital overlap may reduce leading to higher R_b. Compositions 5 and 6 do not follow the same trend and their R_b values seem to be elevated. The curves of R_b vs $\boldsymbol{a}$ for 1 – 6 at alternative temperatures of 20, 50 and 100 K all show equivalent features. According to the phase diagram in Figure 9, compositions 5 and 6 are expected to contain secondary CuO phase, whereas compositions 2 – 4 (and possibly 1) are expected to be Cu-deficient. The presence of extra Cu in 5 and 6 seems to lead to higher bulk CCTO resistance above the trend observed for 1 - 4. This confirms the notion that highly resistive GB regions may contain higher Cu content.

The trend of dielectric permittivity ε_b obtained from equivalent circuit fits vs $\boldsymbol{a}$ is presented in Figure 12b, where dielectric data had been collected at 100 K. The main feature of this data is the increase of dielectric permittivity ε_b with $\boldsymbol{a}$ for compositions 1 – 4, which can not be explained by simple lattice expansion. The Clausius Mossotti equation predicts approximately constant $\varepsilon_b \approx 48$ considering the small changes in ***a*** observed and taking into account the atom polarizabilities obtained from [53]. The secondary phases in 1 – 4, $CaTiO_3$ and/or TiO_2, both have a lower permittivity than CCTO and can not cause such an anomalous increase in ε_b directly, but may well influence the defect mechanism in the CCTO bulk phase responsible for the ε_b increase of up to ≈ 100 above Clausius – Mossotti predictions. In the case that secondary phase CuO would simply extend the existing defect mechanism for 1 – 4 to higher Cu contents, compositions 5 and 6 would be expected to fall onto the same line as 1 – 4, which is clearly not the case, and one or more additional defect mechanisms may well exist. The mechanism detected for 1 – 4 may not involve large differences in Cu content, because in this case a larger variation of R_b with ***a*** would be likely than that detected in Figure 12a.

2. Solid Solution in Cu-Rich Compositions

To elucidate the role of secondary CuO phase on the CCTO bulk resistance, the R_b values for all compositions containing extra CuO (5 – 8) were estimated from the relaxation peak height and peak position in the M'' vs f plots (Figure 11) by using equations (4) and (5).

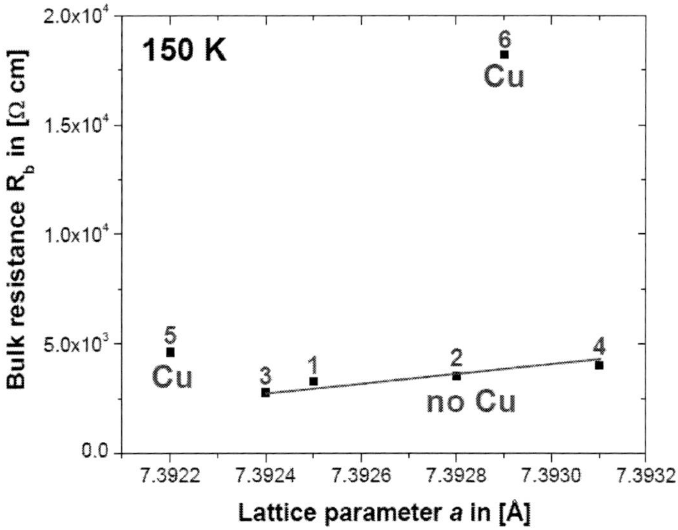

a)

Figure 12. (Continued)

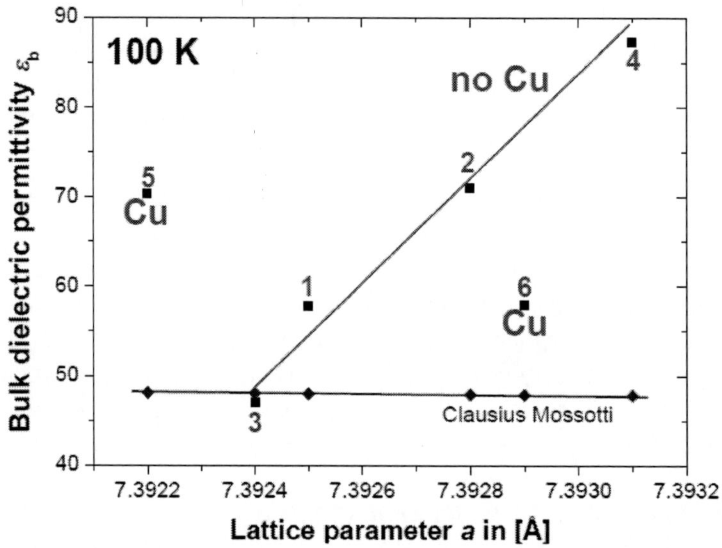

b)

Figure 12. Trends for compositions 1 – 6 of (a) bulk resistance R_b vs lattice parameter *a* (■), and (b) bulk dielectric permittivity ε_b vs *a* (■). Estimations for ε_b in CCTO (♦) were obtained from the Clausius Mossotti equation. R_b and ε_b were obtained from equivalent circuit fits.

For the double-peaks in compositions 7 and 8 estimates were taken from the peak at higher f in the M'' vs f plots, which was argued above to be more likely to represent the CCTO bulk relaxation. The estimated R_b values obtained from M'' vs f plots for 5 and 6 were compared to the respective R_b values from the equivalent circuit fits (section III./C./1.).

It was found that at 80 K the estimated R_b values were smaller by a factor of 3.23 (composition 5) and 3.72 (composition 6), which can be explained by the non-ideality of the relaxation processes leading to a broadening of the relaxation peaks in M'' vs f. This, in turn, leads to a reduction in peak height and an overestimate of ε_b according to equation (5). The error made by the R_b estimates from M'' vs f plots was approximated to be a uniform factor of ≈ 3.5. Thus, the estimated values (□ in Figure 13) for 7 and 8 were multiplied by a factor of 3.5 to account for the estimation error. Both, the uncorrected and corrected R_b values for 7 and 8 at 80 K plotted vs $\boldsymbol{a}$ (■ in Figure 13) approximately fall onto one line with composition 5 and a significant increase of ρ with decreasing $\boldsymbol{a}$ is found. Composition 6 is an exception to this trend with the reasons being unclear.

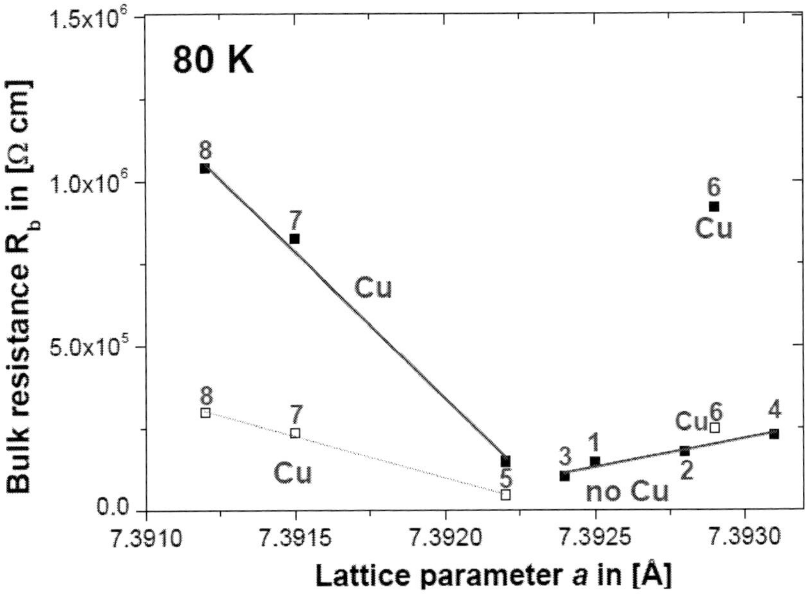

Figure 13. Trends of R_b vs *a* for all compositions (1 – 8) from equivalent circuit fits (■) and from estimates obtained from *M*" vs *f* relaxation peak analysis (□). For compositions 7 and 8 more realistic R_b values (■) were estimated.

Still, the significant R_b increase with decreasing ***a*** detected in Cu-rich samples 5, 7 and 8 strongly suggests the existence of a second defect mechanism and an associated solid solution, which may relate the large differences between GB and bulk resistance to the Cu concentration: GB regions would be expected to be Cu-rich and bulk regions Cu-deficient.

D. Summary and Discussion of the Defect Chemistry in CCTO

Combined analysis of XRD lattice parameter ***a*** and IS data revealed approximately linear relationships between the CCTO bulk resistance R_b vs ***a***, and intrinsic bulk dielectric permittivity ε_b vs ***a***. For Cu-deficient compositions 1 – 4 an anomaleous increase of the CCTO bulk dielectric permittivity ε_b with ***a*** was found above the predictions from the Clausius Mossotti equation. A distinct defect mechanism may well exist causing high intrinsic bulk dielectric permittivity.

Compositions 5 – 8 containing a CuO secondary phase did not follow the same trends. The presence of excess CuO led to increased R_b and indications for a second defect mechanism were found. This second mechanism may explain the large difference between GB and bulk resistance in CCTO in terms of the Cu content. GB regions would be expected to exhibit a higher Cu content as compared to Cu-deficient bulk regions. In compositions 1 – 4 the source of semiconductivity in the respective CCTO bulk phase may well be Cu deficiency too, but it may be independent of the TiO_2 and/or $CaTiO_3$ secondary phases since no strong differences in R_b were detected.

The rather small increase of R_b with $\boldsymbol{a}$ in 1 – 4 (Figure 12a) leads to some further important conclusions: the underlying defect mechanism causing high ε_b may not involve any large changes in the Cu nor the Ti valencies, in which case larger differences in R_b would be likely. This would exclude major contributions from all mechanisms mentioned in section II./C except mechanism (E), which constitutes a Cu excess compensated by the formation of oxygen vacancies. Since compositions 1 – 4 are all Cu deficient, mechanism (E) is unlikely. Instead, the formation of anti-site defects like Ca-Cu, Cu-Ti or Ca-Ti inter-site cation exchanges may be a more plausible explanation. Evidence for Ca-Cu anti-site disorder has indeed been reported in the literature [54], although it should be noted that here in this work the anti-site defects proposed would serve as an explanation for the increased bulk permittivity ε_b and not for the extrinsic giant permittivity ε_{GB}. Furthermore, the possibility still exists that at least one of the defect mechanisms A – D may accompany the anti-site defect formation to a small extent.

The trends of R_b vs $\boldsymbol{a}$ and ε_b vs $\boldsymbol{a}$ are relatively uniform for all compositions investigated which may be a result of the identical powder synthesis and pellet sintering conditions employed. The two distinct defect mechanisms detected were only obvious from combined analysis of $\boldsymbol{a}$ and IS data and cannot be clearly associated to two of the defect mechanisms A - E. From the difficulty of detection it is obvious that the extent to which the defect mechanisms in CCTO ceramics are developed may be small and difficult to control. In the next section it will nevertheless be demonstrated that the extent of both defect mechanisms is directly proportional to the ceramic densification sintering temperature, T_S.

IV. INFLUENCE OF SINTERING TEMPERATURE ON THE IBLC STRUCTURE

A. Dielectric Spectra of CCTO Ceramics Sintered at Various Temperatures

Nominal stoichiometric CCTO powder was synthesized from dried reagents of $CaCO_3$, CuO and TiO_2 at 1000 °C. Six pellets were pressed from fresh CCTO powder and each pellet was sintered at a different temperature T_S, i.e. $T_S = 975$, 1000, 1025, 1050, 1075 and 1100 °C on Pt foil for 12 h followed by slow-cooling in the furnace. For dielectric measurements by IS the pellet surfaces were polished and covered with Au electrodes using dc sputtering. The IS data collected at 80 K for all pellets sintered at different T_S are plotted as the real part of the dielectric permittivity ε' vs frequency f in Figure 14, where two distinct permittivity plateaus are shown representing GB and bulk contributions.

It is apparent that the low-f GB plateau value ε_{GB}* increases substantially with increasing T_S. This increase is far too large to be explained by the changes in grain size (as presented in section IV./C.) or by the increase observed in the pellet density from ≈ 80 to ≈ 95 % for T_S = 975 to 1100 °C. Therefore, it can be concluded that the GB permittivity in CCTO shows a substantial increase with increasing T_S. A magnification of the Figure 14a data is shown in Figure 14b focusing on high f, where the magnitude of the bulk-dominated low permittivity plateau ε_b* increases moderately with T_S as a manifestation of a modest increase of bulk dielectric permittivity ε_b.

As mentioned above in section II./A. the ε_{GB}* and ε_b* plateaus in Figure 14 are composite terms as demonstrated by equations 2 and 3. Such equations are displayed again in Figure 14 now in the permittivity format ε'. Since the R_{GB} and R_b values in the CCTO ceramics investigated here always vary by more than 3 orders of magnitude, ε_{GB}* can be regarded a good estimate for ε_{GB}.

ε_b* is a good estimate for ε_b only in the case that the two capacitors C_{GB} and C_b are sufficiently different from each other. As can be seen in Figure 14a this was not always the case and the contribution from ε_{GB} needs to be considered to calculate ε_b as explained below in section IV./B.

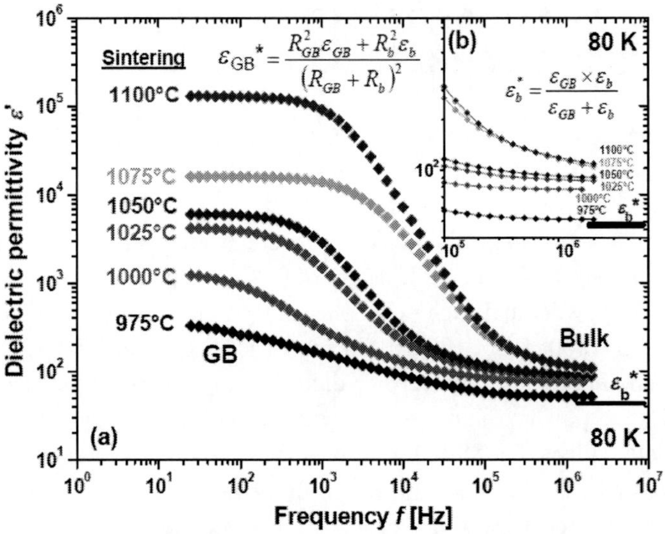

Figure 14. (a) ε' vs f at 80 K for pellets sintered at various T_S. The magnitude of the GB plateau ε_{GB}* increases significantly with T_S. (b) High f behaviour of ε' vs f. Axes are identical to the main figure. The bulk permittivity plateau ε_b* increases with increasing T_S. Reproduced from [52] with permission from the European Ceramic Society.

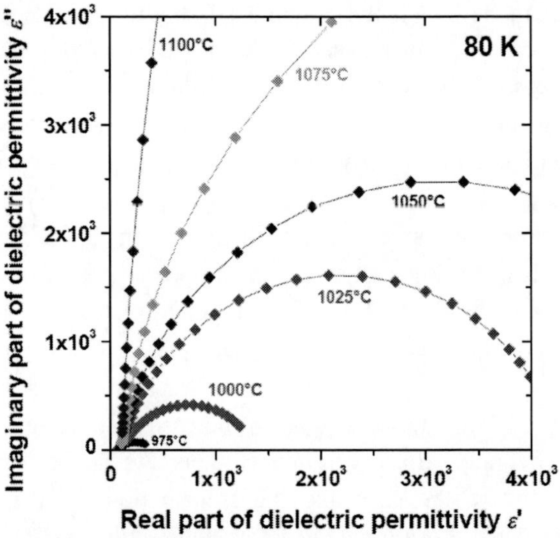

Figure 15. Cole-Cole plots of ε'' vs ε' for CCTO pellets sintered at different temperatures. The diameter of the semicircle is proportional to the GB permittivity. A strong increase of GB permittivity with T_S is evident.

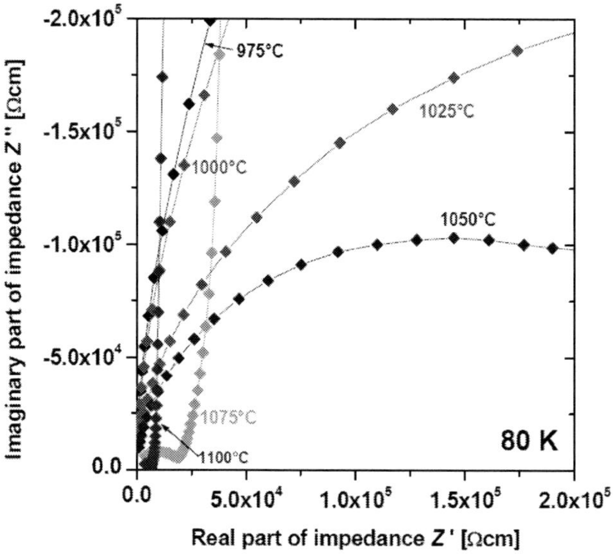

Figure 16. Complex-plane plots of Z'' vs Z' for CCTO pellets sintered at different temperature. The diameter of the high f semicircle is proportional to the bulk resistance R_b. A large decrease of R_b with increasing T_S is observed.

The trend of increasing ε_{GB} with T_S was confirmed from Cole-Cole plots of ε'' vs ε' in Figure 15 displaying a single arc, where the arc diameter is directly proportional to the magnitude of ε_{GB}. ε_{GB}* corresponds to the x-axis (ε' axis) intercept at low f, whereas the high f intercept corresponds to ε_b*, which is close to but not a perfect estimate of ε_b.

A significant decrease of bulk resistance R_b with T_S is indicated in Figure 16 on complex impedance plots of Z'' vs Z', where the bulk semicircle diameter corresponds to R_b. Two semicircles are displayed, which can be seen most clearly for the sample sintered at 1075 °C. At high f an almost fully developed bulk semicircle is shown near the origin of the plot, whereas a second, larger GB semicircle is displayed at lower f. The GB resistance is too large to be fully resolved on this scale and only the onset of the GB semicircle is shown. In fact, the GB resistance was overlaid by a non-ohmic electrode interface relaxation, which did not allow reliable extraction of GB resistance values. For the 1100 °C sintered pellet the bulk semicircle near the origin is not fully developed and a non-zero intercept of the data on the real Z' axis is observed with only the onset of a small bulk semicircle. At 1075 °C the bulk semicircle is clearly visible, whereas for 1050 and 1025 °C sintered ceramics only a fraction of the bulk semicircle can be displayed and the GB semicircle has disappeared at low f due to increased R_{GB}. For 1000 and 975 °C sintered pellets only the onset of the bulk semicircle is shown, emphasizing the large increase in R_b with decreasing T_S.

B. Extraction of Dielectric Parameters and Their Trends with Sintering Temperature

All GB and bulk dielectric parameters were determined quantitatively to more clearly assess their trends with T_S. Equivalent circuit fitting did not succeed for all samples and, therefore, analysis of the IS data was carried out by the following alternative procedure:

(1) GB permittivity ε_{GB} was obtained from the high-permittivity plateau ε_{GB}* at low f in Figure 14a at 80 K. Contributions from the bulk dielectric relaxation were found to be negligible and therefore no correction was made to the ε_{GB}* plateau value.
(2) Bulk permittivity ε_b was obtained from the low permittivity plateau ε_b* at high f from data taken at 50 K, where the ε_1* bulk plateau is fully developed for all samples. In Figure 14b the samples sintered at 1075 and 1100 °C do not show a full relaxation of ε' towards the bulk plateau ε_b* at high f. ε_b* values were corrected and the precise bulk permittivity ε_b obtained by taking into account a small contribution from the GB permittivity ε_{GB}, according to equation (3).
(3) Bulk resistance R_b was calculated from the bulk permittivity ε_b and the frequency of the bulk dielectric relaxation peaks displayed in plots of the electric modulus function M'' vs f (data not shown here), according to equation (4).

All trends of GB permittivity ε_{GB}, bulk permittivity ε_b and bulk resistance R_b vs T_S at selected temperatures are shown in Figure 17a-c. The ε_{GB} values show an exponential increase with T_S by a factor of ≈ 300 (Figure 17a), which is difficult to interpret. It could reflect intrinsic changes in the GB regions or a geometrical effect associated with GB regions getting thinner with increasing T_S. The detected ε_{GB} increase is in agreement with previous work on CCTO ceramics sintered at different T_S [55].

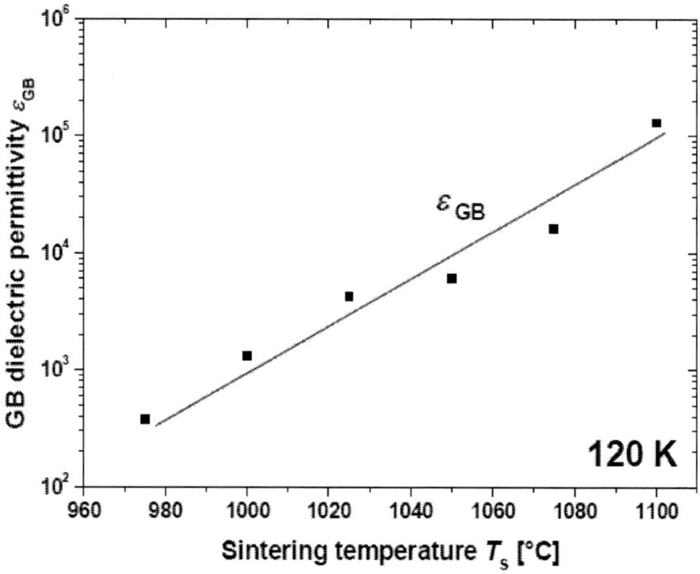

a)

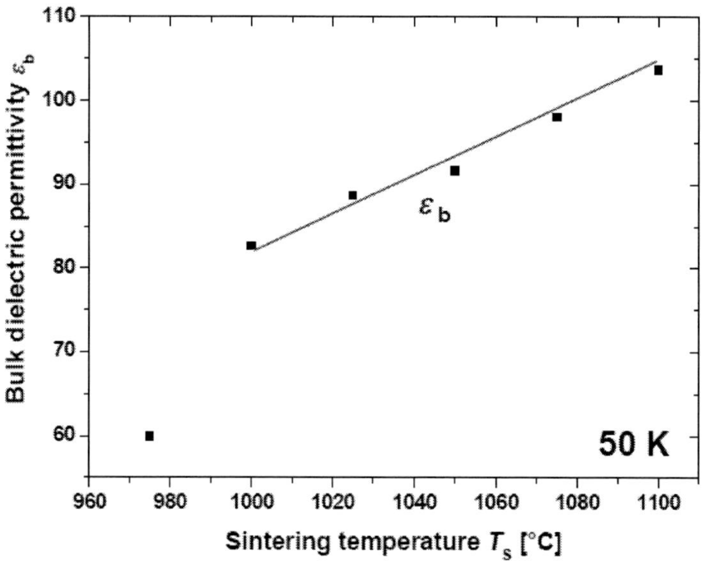

b)

Figure 17. (Continued)

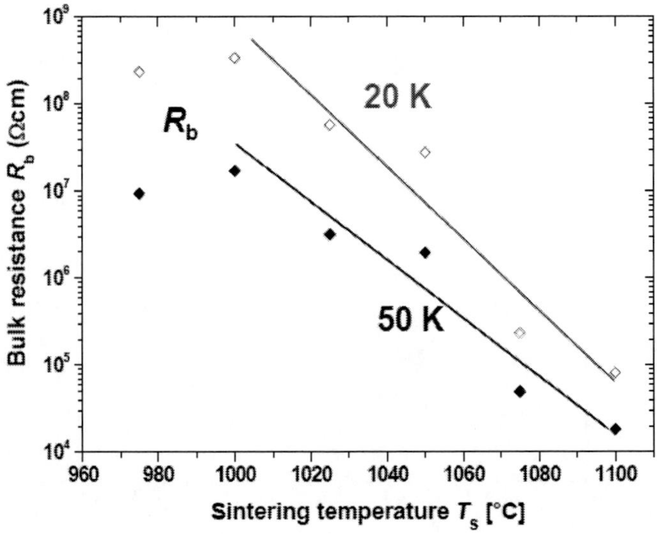

c)

Figure 17. (a) ε_{GB} vs T_S at 120 K. An exponential increase of ε_{GB} with T_S is shown. (b) ε_b vs T_S at 50 K. A linear increase of ε_b with T_S is shown. (c) R_b vs T_S at 20 (◊) and 50 K (♦). An exponential decrease of R_b with T_S is shown. Solid lines are guide to the eyes. Images reproduced from [52] with permission from the European Ceramic Society.

This increase is significantly more pronounced here as was found in previous studies on 1100 °C sintered CCTO with different sintering time [56, 57]. T_S as opposed to sintering time has a more pronounced influence on the dielectric properties of CCTO, which may reflect the importance of diffusion-related phenomena occurring during the densification sintering process in CCTO ceramics. The intrinsic bulk permittivity ε_b shows an approximately linear increase with T_S between 1000 - 1100 °C by a factor of ≈ 2 (Figure 17b), which can be associated with intrinsic changes of the bulk phase. Only a small part of this increase can be associated with changes in pellet density.

The rather high intrinsic ε_b values of ≈ 100 in CCTO and related materials have been associated previously with the formation of an incipient ferroelectric phase [58]. The formation of this phase seems to be particularly pronounced here by increasing T_S from 975 to 1000 °C. As mentioned in section III./C./1., the formation of the incipient ferroelectric phase may possibly be associated with a defect mechanism based on Ca-Cu anti-site defects. Apparently, this mechanism is strongly influenced by T_S leading to a linear increase of ε_b with increasing T_S and anti-site disorder would gradually evolve with T_S to higher levels.

In Figure 17c the bulk resistance R_b shows an exponential decrease with increasing T_S between 1000 – 1100 °C by a factor of ≈ 1000 – 4000, but no decrease can be observed by increasing T_S from 975 to 1000 °C. Since a clear increase of ε_b was shown in Figure 17b from 975 to 1000 °C, the ε_b increase with T_S may rely on a different mechanism in agreement with the arguments proposed in section III./C./2. It is likely that the defect mechanism responsible for the massive reduction of R_b with T_S may involve the Cu concentration as mentioned in section III./C./2. This is confirmed by the notion that increasing Cu segregation out of the CCTO ceramics is the only visible change of the CCTO phase with increasing T_S as will be demonstrated in the next section.

C. Influence of Sintering Temperature on the CCTO Composition and Surface Morphology

Scanning electron microscopy (SEM) combined with energy dispersive analysis of X-rays (EDAX) were carried out on all pellets sintered at different T_S to investigate any changes in chemical composition and ceramic microstructure. High resolution quantitative EDAX line scans were carried out using a narrow detector opening to achieve maximum spatial resolution along the single scanning line on the pellet surfaces to demonstrate the development of a Cu-rich segregated inter-granular phase with increasing T_S. In an attempt to detect Cu non-stoichiometry in the main CCTO bulk phase, high resolution quantitative EDAX spot analysis on the centre of single CCTO grains was carried out. For each sintered pellet the Ca:Cu:Ti ratios on the centre of 25 different grains were measured to obtain a statistical average for each sample. The penetration depth of the electron beam spot on one single CCTO grain was estimated to be 1 – 2 μm and, therefore, the main X-ray intensity detected was expected to be emitted from the CCTO grain centres irrespective of the inter-granular phase. The quantitative EDAX Ca:Cu:Ti ratios were corrected with respect to standard samples consisting in a CuTi alloy and a $CaTiO_3$ ceramic with predetermined 1:1 Cu:Ti and Ca:Ti atomic ratios. Furthermore, several large area EDAX scans (≈ 150 × 150 μm) with a wide detector opening were carried out on each pellet to obtain the average surface cation concentrations. The mean grain sizes were determined from the respective backscattered SEM images by averaging the size of 220 randomly selected

CCTO grains on each sample and the grain size distributions were investigated.

SEM images in the backscattered electron mode and EDAX line scans on the unpolished surfaces of CCTO ceramics sintered at 1025 and 1075 °C are shown in Figure 18. EDAX data were collected from left to right along the solid (false-coloured orange) lines to determine the cation concentrations. In the backscattered mode the image colour contrast emphasizes differences in atomic number and is therefore ideal to detect the segregated Cu-rich secondary phase in CCTO. From this analysis, a Cu-rich phase starts to segregate out of the ceramics at $T_S \approx 1050$ °C and accumulates between the CCTO grains as an inter-granular phase. The segregated phase is clearly seen at 1075 °C whereas there are no detectable signs of it for $T_S = 1025$ °C.

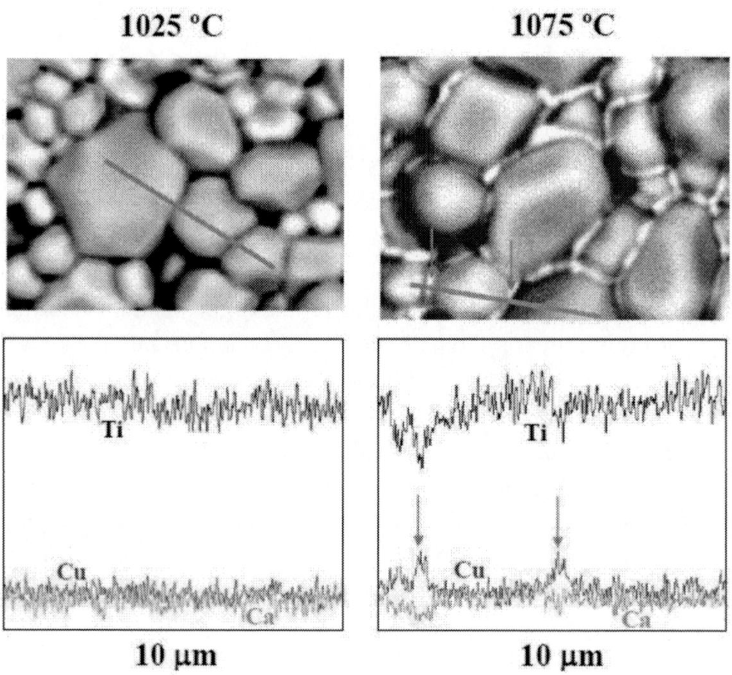

Figure 18. SEM images collected in backscattered electron mode from unpolished surfaces of CCTO pellets sintered at 1025 and 1075 °C (first row), and EDAX line scans (second row). The solid (orange) lines in the first row indicate the quantitative EDAX line scans of ≈ 10 μm, with the Ti, Cu, Ca and oxygen atomic concentrations along these lines being presented in the second row. Cu segregation at $T_S = 1075$ °C in form of an inter-granular phase is indicated. Reproduced from [52] with permission from the European Ceramic Society.

EDAX point measurements for the 1075 °C sintered pellet indicated a high Cu content of the inter-granular phase (about 68 %). The remaining 32% of Ca and Ti cations may well be a proximity effect due to the adjacent CCTO phase and the segregated phase is therefore most likely copper oxide and not a Cu-rich CCTO type phase. This inter-granular Cu_xO phase could not be detected by XRD, probably due to its small volume fraction, or it may be amorphous and no XRD detection was therefore possible. It should be noted that the detected inter-granular phase does not fully cover all CCTO grains and can not serve as an explanation for the large difference in dielectric properties between GB and bulk relaxations in CCTO. On the other hand, it looks quite likely that GB regions in CCTO would be Cu-rich due to the detected Cu segregation towards the GBs and Cu accumulation as an inter-granular phase.

All XRD patterns collected showed evidence for only the CCTO phase to be present up to T_S = 1100 °C but phase decomposition was indicated above T_S = 1125 °C. For T_S = 1125 °C the XRD pattern showed the presence of $CaTiO_3$, TiO_2 and small residues of the main CCTO phase, but no Cu_xO phase. Furthermore, no signs of Cu_xO phase were detected in the SEM backscattered images either after sintering at 1125 °C. This may be understood within the following scenario: Cu_xO segregation starts to occur above T_S = 1050 °C leading to a Cu-deficient CCTO bulk phase and a range of Cu-deficient compositions may exist, constituting one or more defect mechanisms in the form of small solid solutions as mentioned in section III./C./2. Cu segregation occurs towards the GB regions, which may therefore contain a higher Cu-content and display higher resistance in agreement with the arguments proposed in section III./C./2. At higher T_S Cu segregation may intensify and the increasing Cu-deficiency in the CCTO bulk may ultimately lead to the decomposition of the CCTO phase, where the segregated Cu_xO may volatilize. Therefore, the volatility of Cu within the CCTO structure and its segregation towards GBs may be regarded the main driving force for the formation of the IBLC structure in CCTO ceramics.

Figure 19 shows the results from high resolution, quantitative EDAX spot analysis of single CCTO ceramic grains, corrected to standard compounds. The Cu contents in terms of the atomic fractions for an average of 25 grains and for the large area average (both for unpolished sintered pellets) are displayed vs T_S. The solid horizontal (green) line represents the expected 0.375 Cu cationic ratio (atomic fraction). Cu segregation is clearly evident due to Cu-excess on the unpolished surface above $T_S \geq$ 1050 °C.

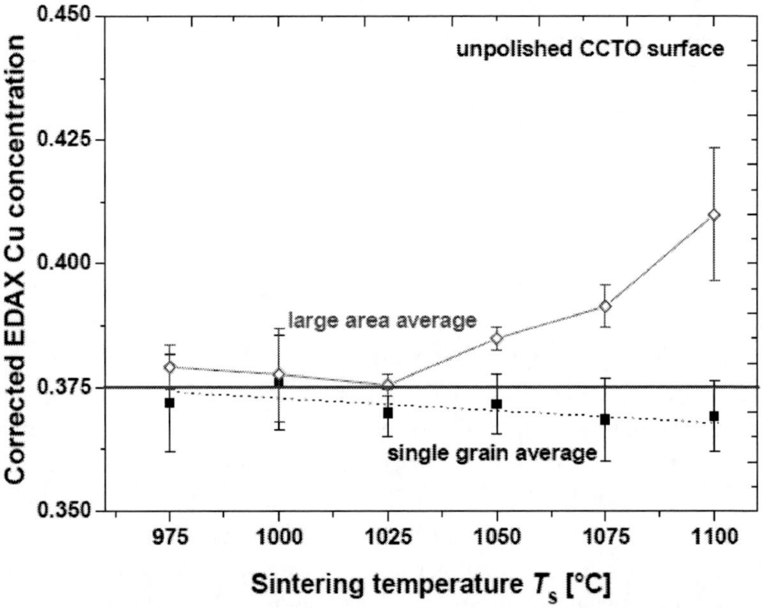

Figure 19. Quantitative EDAX results corrected to Ca, Cu and Ti standards. Data were obtained from an average of 25 single grains in each pellet (unpolished) (■) and from large area average scans (5 large scans of 150 × 150 μm for each sample) of the same pellets (◊). Image reproduced from [52] with permission from the European Ceramic Society.

Averaging the Cu content in 25 single grains gave a weak trend towards Cu-loss from within the grain interiors but the trend is statistically just on the limit of significance due to the large experimental error. The error bars shown represent the first standard deviation of the Cu concentration in the 25 grains investigated, which constitutes a conservative error estimate.

Considering that the data points for single grains in Figure 19 represent the highest probability of the actual Cu content of the CCTO grains and that the error estimate is conservative, it does not seem plausible that the clear trend observed on 6 data points is a statistical coincidence. Large area scans taken from polished surfaces showed no Cu accumulation. Obviously, the Cu-rich phase accumulates at the pellet surfaces during sintering and can be removed by polishing. From the backscattered SEM images shown in Figure 20 the mean grain sizes were determined from 220 randomly selected CCTO grains, where the grain size distributions at different T_S shown in Figure 20 indicate a moderate grain growth with increasing T_S.

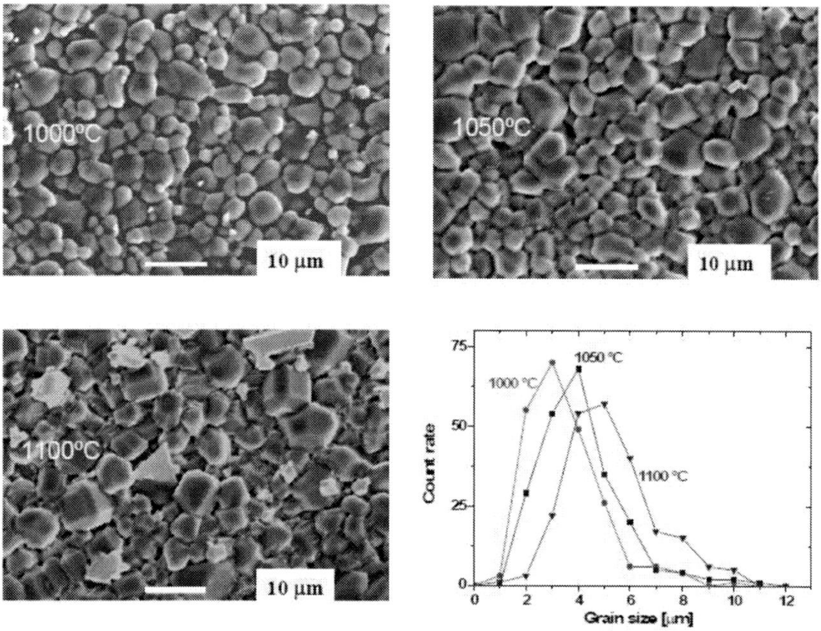

Figure 20. SEM images collected with secondary electrons for CCTO pellets sintered at different temperatures, T_S. Moderate grain growth is evident with increasing T_S. The grain size distributions in the bottom right panel confirm such grain growth.

The average grain size increased approximately linearly from ≈ 3 μm for T_S = 1000 °C to ≈ 4.8 μm for T_S = 1100 °C. The grain growth is accompanied by an increase in pellet density as expected for a densification sintering processes from ≈ 80% for T_S = 1000 °C to ≈ 95% for T_S = 1100 °C. The moderate grain growth detected is insufficient to explain the increase in ε_{GB} displayed in Figure 14a and 17a as mentioned above.

D. Summary of the Effects of Sintering Temperature on CCTO Ceramics

Increasing the sintering temperature of CCTO ceramics from 975 to 1100 °C leads to a substantial variation in the dielectric properties. The GB permittivity increases exponentially by a factor of ≈ 300, the bulk permittivity increases linearly by a factor of ≈ 2, and the bulk resistance decreases exponentially by a factor of ≈ 10^3. The extent of the CCTO defect mechanisms is strongly dependent on T_S and evolves significantly with increasing T_S. The bulk semiconductivity is most likely dependent on the Cu content, which was confirmed from quantitative EDAX analysis indicating Cu-loss within individual CCTO grains with increasing T_S.

Conclusion and Discussion

CCTO ceramics consist of insulating GB and conducting bulk regions where electronic heterogeneity may be explained by the defect chemistry of the compound. At least two different defect mechanisms exist which may explain (A) resistive GBs and semiconducting bulk by variations in Cu content, and (B) high bulk dielectric permittivity ε_b in the range of $\approx$ 100 possibly due to Ca-Cu anti-site defects. The defect mechanisms both develop during the sintering densification process in CCTO ceramics, where variations in Cu content come about due to Cu segregation towards GB regions.

The CCTO giant permittivity values of up to 300 000 reported before are extrinsic in origin, which poses certain limitations for application of CCTO as a capacitor material. Several reports in the literature show that the thin insulating GBs in CCTO exhibit non-linear resistance and may be understood in terms of Schottky-type barriers [3, 16], which is generally disadvantageous in a charge storage device. However, the non-linearity is visible by macroscopic measurement techniques only for CCTO ceramics with very large grain size in the range of 100 μm or more, whereas the CCTO ceramics discussed here showed linear resistance up to the experimental limitation of 20 V. CCTO ceramics with grain sizes in the range of a few micrometers may therefore be good candidates for capacitor applications due to their high permittivity as a result of a core-shell structure of insulating GBs and conducting bulk. Large grain CCTO ceramics on the other hand with strongly non-linear resistance may be interesting to be considered for varistor applications.

Dedication

The authors wish to dedicate this review to the memory of the late Vito Raineri († 18.2.2012) who led the NUOTO project and is sadly missed by all his colleagues.

ACKNOWLEDGMENTS

The authors wish to acknowledge funding from the European Union (EU) under the NUOTO project and under the FP7/2007-2013 grant agreement n° 226716.

REFERENCES

[1] Subramanian, M. A.; Li, D.; Duan, N.; Reisner, B. A.; Sleight, A. W. High Dielectric Constant in $ACaCu_3Ti_4O_{12}$ and $ACu_3Ti_3FeO_{12}$ Phases. *J. Solid State Chem.* 2000, vol. 151, p.323.

[2] Homes, C. C.; Vogt, T.; Shapiro, S. M.; Wakimoto, S.; Ramirez, A. P. Optical Response of High-Dielectric-Constant Perovskite-Related Oxide. *Science* 2001, vol. 293, p.673.

[3] Chung, S.-Y.; Kim, I.-D.; Kang, S.-J. L. Strong nonlinear current-voltage behaviour in perovskite-derivative calcium copper titanate. *Nat. Mater.* 2004, vol. 3, p.774.

[4] Adams, T. B.; Sinclair, D. C.; West, A. R. Giant Barrier Layer Capacitance Effects in $CaCu_3Ti_4O_{12}$ Ceramics. *Adv. Mater.* 2002, vol. 14, p.1321.

[5] Schmidt, R.; Sinclair, D. C. Anomalous increase of dielectric permittivity in Sr-doped CCTO ceramics $Ca_{1-x}Sr_xCu_3Ti_4O_{12}$ ($0 \leq x \leq 0.2$). *Chem. Mater.* 2010, vol. 22, p.6.

[6] Brizé, V.; Gruener, G.; Wolfman, J.; Fatyeyeva, K.; Tabellout, M.; Gervais, M.; Gervais, F. Grain size effects on the dielectric constant of $CaCu_3Ti_4O_{12}$ ceramics. *Mater. Sci. Eng. B* 2006, vol. 129, p.135.

[7] West, A. R.; Adams, T. B.; Morrison, F. D.; Sinclair, D. C. Novel high capacitance materials:- $BaTiO_3$:La and $CaCu_3Ti_4O_{12}$. *J. Eur. Ceram. Soc.* 2004, vol. 24, p.1439.

[8] Deng, G.; Yamada, T.; Muralt, P. Evidence for the existence of a metal-insulator-semiconductor junction at the electrode interfaces of $CaCu_3Ti_4O_{12}$ thin film capacitors. *Appl. Phys. Lett.* 2007, vol. 91, p.202903.

[9] Fiorenza, P.; Nigro, R. L.; Bongiorno, C.; Raineri, V.; Ferarrelli, M. C.; Sinclair, D. C.; West, A. R. Localized electrical characterization of the giant permittivity effect in $CaCu_3Ti_4O_{12}$ ceramics. *Appl. Phys. Lett.* 2008, vol. 92, p.182907.

[10] Si, W.; Cruz, E. M.; Johnson, P. D.; Barnes, P. W.; Woodward, P.; Ramirez, A. P. Epitaxial thin films of the giant-dielectric-constant material $CaCu_3Ti_4O_{12}$ grown by pulsed-laser deposition. *Appl. Phys. Lett.* 2002, vol. 81, p.2056.

[11] Deng, G.; He, Z.; Muralt, P. Physical aspects of colossal dielectric constant material $CaCu_3Ti_4O_{12}$ thin films. *J. Appl. Phys.* 2009, vol. 105, p.084106.

[12] Kant, C.; Rudolf, T.; Mayr, F.; Krohns, S.; Lunkenheimer, P.; Ebbinghaus, S. G.; Loidl, A. Broadband dielectric response of $CaCu_3Ti_4O_{12}$: From dc to the electronic transition regime. *Phys. Rev. B* 2008, vol. 77, p.045131.

[13] Li, M.; Shen, Z.; Nygren, M.; Feteira, A.; Sinclair, D. C.; West, A. R. Origin(s) of the apparent high permittivity in $CaCu_3Ti_4O_{12}$ ceramics: clarification on the contributions from internal barrier layer capacitor and sample-electrode contact effects. *J. Appl. Phys.* 2009, vol. 106, p.104106.

[14] Ferrarelli, M. C.; Sinclair, D. C.; West, A. R.; Dabkowska, H. A.; Dabkowski, A.; Luke, G. M. Comment on the origin(s) of the giant permittivity effect in $CaCu_3Ti_4O_{12}$ single crystals and ceramics. *J. Mater. Chem.* 2009, vol. 19, p.5916.

[15] Sinclair, D. C.; Adams, T. B.; Morrison, F. D.; West, A. R. $CaCu_3Ti_4O_{12}$: One-step internal barrier layer capacitor. *Appl. Phys. Lett.* 2002, vol. 80, p.2153.

[16] Adams, T. B.; Sinclair, D. C.; West, A. R. Characterization of grain boundary impedances in fine- and coarse-grained $CaCu_3Ti_4O_{12}$ ceramics. *Phys. Rev. B* 2006, vol. 73, p.094124.

[17] Li, M.; Sinclair, D. C.; West, A. R. Extrinsic origins of the apparent relaxorlike behavior in $CaCu_3Ti_4O_{12}$ ceramics at high temperatures: A cautionary tale. *J. Appl. Phys.* 2011, vol. 109, p.084106.

[18] Lunkenheimer, P.; Fichtl, R.; Ebbinghaus, S. G.; Loidl, A. Nonintrinsic origin of the colossal dielectric constants in $CaCu_3Ti_4O_{12}$. *Phys. Rev. B* 2004, vol. 70, p.172102.

[19] Kishi, H.; Okino, Y.; Honda, M.; Iguchi, Y.; Imaeda, M.; Takahashi, Y.; Ohsato, H.; Okuda, T. The Effect of MgO and Rare-Earth Oxide on Formation Behavior of Core-Shell Structure in $BaTiO_3$. *Japanese J. Appl. Phys.* 1997, vol. 36, p.5954.

[20] Shannon, R. D.; Prewitt, C. T. Effective Ionic Radii in Oxides and Fluorides. *Acta Cryst. B* 1969, vol. 25, p.925.

[21] Barsukov, E.; Macdonald, J. *Impedance Spectroscopy: Theory, Experiment and Applications*. John Wiley and Sons Inc., Hoboken, 2005.

[22] Schmidt, R., In: *Ceramic Materials Research Trends*, edited by Lin, P. B. Novascience Publishers, Hauppauge, 2007, p.321.

[23] Irvine, J. T. S.; Sinclair, D. C.; West, A. R. Electroceramics: Characterization by Impedance Spectroscopy. *Adv. Mater.* 1990, vol. 2, p.132.

[24] Hsu, C. H.; Mansfeld, F. Technical Note: Concerning the Conversion of the Constant Phase Element Parameter Y_0 into a Capacitance. *Corrosion* 2001, vol. 57, p.747.

[25] Krohns, S.; Lunkenheimer, P.; Ebbinghaus, S. G.; Loidl, A. Broadband dielectric spectroscopy on single-crystalline and ceramic $CaCu_3Ti_4O_{12}$. *Appl. Phys. Lett.* 2007, vol. 91, p.022910.

[26] Capsoni, D.; Bini, M.; Massarotti, V.; Chiodelli, G.; Mozzatic, M. C.; Azzoni, C. B. Role of doping and CuO segregation in improving the giant permittivity of $CaCu_3Ti_4O_{12}$. *J. Solid State Chem.* 2004, vol. 177, p.4494.

[27] Li, M. in *Department of Engineering Materials* The University of Sheffield, Sheffield, 2008, PhD.

[28] Li, M.; Feteira, A.; Sinclair, D. C.; West, A. R. Influence of Mn doping on the semiconducting properties of $CaCu_3Ti_4O_{12}$ ceramics. *Appl. Phys. Lett.* 2006, vol. 88, p.232903.

[29] Li, J.; Sleight, A. W.; Subramanian, M. A. Evidence for internal resistive barriers in a crystal of the giant dielectric constant material: $CaCu_3Ti_4O_{12}$. *Solid State Commun.* 2005, vol. 135, p.260.

[30] Fang, T.-T.; Mei, L.-T.; Ho, H.-F. Effects of Cu stoichiometry on the microstructures, barrier-layer structures, electrical conduction, dielectric responses, and stability of $CaCu_3Ti_4O_{12}$. *Acta Materialia* 2006, vol. 54, p.2867.

[31] Deng, G.; Xanthopoulos, N.; Muralt, P. Chemical nature of colossal dielectric constant of $CaCu_3Ti_4O_{12}$ thin film by pulsed laser deposition. *Appl. Phys. Lett.* 2008, vol. 92, p.172909.

[32] Fang, T.-T.; Mei, L.-T. Evidence of Cu Deficiency: A Key Point for the Understanding of the Mystery of the Giant Dielectric Constant in $CaCu_3Ti_4O_{12}$. *J. Am. Ceram. Soc.* 2007, vol. 90, p.638.

[33] Kim, C. H.; Jang, Y. H.; Seo, S. J.; Song, C. H.; Son, J. Y.; Yang, Y. S.; Cho, J. H. Effect of Mn doping on the temperature-dependent anomalous

giant dielectric behavior of $CaCu_3Ti_4O_{12}$. *Phys. Rev. B* 2012, vol. 85, p.245210.

[34] Schmidt, R.; Ventura, J.; Langenberg, E.; Nemes, N. M.; Munuera, C.; Varela, M.; Garcia-Hernandez, M.; Leon, C.; Santamaria, J. Magnetoimpedance spectroscopy of epitaxial multiferroic thin films. *Phys. Rev. B* 2012, vol. 86, p.035113.

[35] Ferrarelli, M. C.; Adams, T. B.; Feteira, A.; Sinclair, D. C.; West, A. R. High intrinsic permittivity in $Na_{1/2}Bi_{1/2}Cu_3Ti_4O_{12}$. *Appl. Phys. Lett.* 2006, vol. 89, p.212904.

[36] Hong, S.-H.; Kim, D.-Y.; Park, H.-M.; Kim, Y.-M. Electric and Dielectric Properties of Nb-Doped $CaCu_3Ti_4O_{12}$ Ceramics. *J. Am. Ceram. Soc.* 2007, vol. 90, p.2118.

[37] Liu, J.; Duan, C.-G.; Mei, W. N.; Smith, R. W.; Hardy, J. R. Dielectric properties and Maxwell-Wagner relaxation of compounds $ACu_3Ti_4O_{12}$ ($A = Ca,Bi_{2/3},Y_{2/3},La_{2/3}$). *J. Appl. Phys.* 2005, vol. 98, p.093703.

[38] Sebald, J.; Krohns, S.; Lunkenheimer, P.; Ebbinghaus, S. G.; Riegg, S.; Reller, A.; Loidl, A. Colossal dielectric constants: A common phenomenon in $CaCu_3Ti_4O_{12}$ related materials. *Solid State Commun.* 2010, vol. 150, p.857.

[39] Feng, L.; Tang, X.; Yan, Y.; Chen, X.; Jiao, Z.; Cao, G. Decrease of dielectric loss in $CaCu_3Ti_4O_{12}$ ceramics by La doping. *physica status solidi (a)* 2006, vol. 203, p.R22.

[40] Hu, Y.; Jeng, T.-S.; Liu, J.-S. Effect of the MgO substitution for CuO on the properties of $CaCu_3Ti_4O_{12}$ ceramics. *Ceram. Intern.* 2012, vol. 38, p.3459.

[41] Rai, A. K.; Singh, N. K.; Acharya, S. K.; Singh, L.; Mandal, K. D. Effect of tantalum substitutions on microstructures and dielectric properties of calcium copper titanate ($CaCu_3Ti_4O_{12}$) ceramic. *Mater. Sci. Eng. B* 2012, vol. 177, p.1207.

[42] Li, W. L.; Zhao, Y.; Chi, Q. G.; Zhang, Z. G.; Fei, W. D. Enhanced performance of core-shell-like structure Zr-doped $CaCu_3Ti_4O_{12}$ ceramics prepared by a flame synthetic approach. *RSC Advances* 2012, vol. 2, p.6073.

[43] Zheng, X. H.; Zhang, C.; Liang, B. L.; Tang, D. P.; Huang, X.; Liu, X. L. An effective way to tune the microstructure and dielectric properties of $CaCu_3Ti_4O_{12}$ ceramics. *J. Alloys Comp.* 2010, vol. 505, p.L10.

[44] Zhang, L.; Wu, Y.; Guo, X.; Wang, Z.; Zou, Y. Influence of Zr doping on the dielectric properties of $CaCu_3Ti_4O_{12}$ Ceramics. *J. Mater. Sci. Mater. Electron.* 2012, vol. 23, p.865.

[45] Kwon, S.; Huang, C.-C.; Patterson, E. A.; Cann, D. P.; Alberta, E. F.; Kwon, S.; Hackenberger, W. S.; Cann, D. P. The effect of Cr_2O_3, Nb_2O_5 and ZrO_2 doping on the dielectric properties of $CaCu_3Ti_4O_{12}$. *Mater. Lett.* 2008, vol. 62, p.633.

[46] Choi, S.-W.; Hong, S.-H.; Kim, Y.-M. Effect of Al Doping on the Electric and Dielectric Properties of $CaCu_3Ti_4O_{12}$. *J. Am. Ceram. Soc.* 2007, vol. 90, p.4009.

[47] Ferrarelli, M. C. in *The Department of Engineering Materials* The University of Sheffield, Sheffield, 2008, PhD.

[48] Scher, H.; Zallen, R. Critical Density in Percolation Processes. *J. Chem. Phys.* 1970, vol. 53, p.3759.

[49] Vertruyen, B.; Cloots, R.; Ausloos, M.; Fagnard, J.-J.; Vanderbemden, P. Electrical transport and percolation in magnetoresistive manganite/insulating oxide composites: Case of $La_{0.7}Ca_{0.3}MnO_3/Mn_3O_4$. *Phys. Rev. B* 2007, vol. 75, p.165112.

[50] Veith, M.; Ren, S.; Wittmar, M.; Bolz, H. Giant dielectric constant response of the composites in ternary system CuO-TiO_2-CaO. *J. Solid State Chem.* 2009, vol. 182, p.2930.

[51] Abram, E. J.; Sinclair, D. C.; West, A. R. A Strategy for Analysis and Modelling of Impedance Spectroscopy Data of Electroceramics: Doped Lanthanum Gallate. *J. Electroceram.* 2003, vol. 10, p.165.

[52] Schmidt, R.; Stennett, M. C.; Hyatt, N. C.; Pokorny, J.; Prado-Gonjal, J. s.; Li, M.; Sinclair, D. C. Effects of sintering temperature on the internal barrier layer capacitor (IBLC) structure in $CaCu_3Ti_4O_{12}$ (CCTO) ceramics. *J. Eur. Ceram. Soc.* 2012, vol. 32, p.3313.

[53] Shannon, R. D. Dielectric polarizabilities of ions in oxides and fluorides. *J. Appl. Phys.* 1993, vol. 73, p.348.

[54] Zhu, Y.; Zheng, J. C.; Wu, L.; Frenkel, A. I.; Hanson, J.; Northrup, P.; Ku, W. Nanoscale Disorder in $CaCu_3Ti_4O_{12}$: A New Route to the Enhanced Dielectric Response. *Phys. Rev. Lett.* 2007, vol. 99, p.037602.

[55] Bender, B.; Pan, M. The effect of processing on the giant dielectric properties of $CaCu_3Ti_4O_{12}$. *Mater. Sci. Eng. B* 2005, vol. 117, p.339.

[56] Romero, J. J.; Leret, P.; Rubio-Marcos, F.; Quesada, A.; Fernández, J. F. Evolution of the intergranular phase during sintering of $CaCu_3Ti_4O_{12}$ ceramics. *J. Eur. Ceram. Soc.* 2010, vol. 30, p.737.

[57] Shri Prakash, B.; Varma, K. B. R. Effect of sintering conditions on the dielectric properties of $CaCu_3Ti_4O_{12}$ and $La_{2/3}Cu_3Ti_4O_{12}$ ceramics: A comparative study. *Physica B: Cond. Mat.* 2006, vol. 382, p.312.

[58] Ferrarelli, M. C.; Nuzhnyy, D.; Sinclair, D. C.; Kamba, S. Soft-mode behavior and incipient ferroelectricity in $Na_{1/2}Bi_{1/2}Cu_3Ti_4O_{12}$. *Phys. Rev. B* 2010, vol. 81, p.224112.

In: Capacitors
Editor: Kristofer N. Muller

ISBN: 978-1-62417-585-5

Chapter 2

DEVICES, MATERIALS AND MECHANISMS OF CAPACITIVE SENSORS FOR GAS AND HUMIDITY DETECTIONS

***Chi Lu*[1,*] *and Ping Yu*[2]**

[1]Department of Electrical and Computer Engineering and Center for Nanoscale Science and Engineering, University of Kentucky, Lexington, KY, US

[2]Metabolism Branch, National Cancer Institute, National Institutes of Health, Bethesda, US

ABSTRACT

Capacitive gas sensors are a distinct group of microelectronic devices that detect the analyte concentration by the change of capacitance. While currently resistor-based devices dominate the domain of solid-state gas detection, specific advantages allow gas sensing capacitors to be favored in many circumstances with promising prospects. Starting with an overview of gas detection, this chapter reviews the history and recent progresses of several kinds of capacitive gas or humidity sensors. Detailed discussions are carried out on the device compositions, sensing materials, and principles. Dielectric or electrode materials used and studied in different capacitive gas sensors including organic, inorganic,

* E-mail: chi_lu_uky@yahoo.com.

and novel nano-scale structures, as well as proved or proposed detection processes, especially the reversible interactions between the analyte and the sensing materials, are described along with the corresponding type of devices. Some non-capacitive solid-state or microelectronic sensors with identical or alike sensing mechanisms to the gas sensing capacitors will also be discussed. Besides, sensor design criteria and device fabrications are briefly introduced.

Keywords: Capacitive sensor, gas detection, microelectronic devices, dielectric, sensing material, mechanism

1. An Overview of Gas Detection

Gas detection is a primary task in human society. There are explosive gases including hydrogen and methane, toxic gases such as carbon monoxide and chlorine, pollutants like nitrogen oxides, and harmless but essential gaseous substances, e. g., water vapor, oxygen, carbon dioxide, etc. All these gases, no matter from the natural environment or human activities, need to be monitored for the well-being of mankind. Based on gas sensors in commercial markets and scientific literatures, several detection methodologies and numerous sensing materials have been developed. The sensing devices can be roughly divided into two fundamental categories: solid-state and non-solid-state [1]. Non-solid-state detectors, though possess improved accuracies, are usually bulky and not quite adaptable to real-time monitoring. Portability, immediate response and low cost are foremost merits of solid-state devices that make up the mainstream of commercial and state-of-the-art gas detectors.

Four major types of solid-state gas sensors prevail in both market and research: optic, mass, electrochemical, and electronic. Optic sensors typically use the changes of incident visible lights caused by responses of sensing materials to the analyte gas [2]. The variation of the characteristics of a certain surface acoustic wave (SAW) resulting from mass change of the sensing material is widely adopted in mass detectors [3]. Mass change can also alter the resonant frequency of a quartz crystal microbalance (QCM) that serves as another typical approach in mass sensors [4]. A solid-state electrolyte is the key part for most of electrochemical gas sensors, and therefore the analyte or the final product stemmed from the analyte must be ionized [5].

In this chapter, an "electronic sensor" is defined by that the signal results directly from the reversible change of certain electronic properties of the sensing material, such as resistance (or conductance), capacitance, inductance, or any field-effect characteristics. Currently most electronic sensors are fabricated through micro- and nanotechnologies. The sizes of key components of devices, including one- (e.g., nanowires), two- (thin films) and three-dimensional small structures, are usually in the scale of micrometers or even nanometers [6]. Therefore, this type of devices are also called "microelectronic sensors".

The majority of reported microelectronic gas sensors are based on the resistance change. Great progresses have been made on MOS (metal-oxide-semiconductor) devices with signals derived from the capacitance-voltage (C-V) correlation during the last thirty years [7]. Non-MOS devices based on the variation of capacitance resulting from the analyte, or "capacitive sensors", are not as popular as the former two. However, boasting several merits, such as immediate response, miniaturization, compatibility to microelectronic manufacturing, low cost, low power-consumption, and simple signal treatment (theoretically only an oscillator circuit is required), capacitive sensors has good prospects in detecting gases and humidity. Also, the fundamental structure of this type of devices requires only three constituents, two conductive coatings or contacts and one dielectric film sandwiched in between. Therefore, the manufacturing complexity of gas sensors based on simple capacitors is expected to be much less than many other devices. Moreover, a capacitive sensor is able to be included in a passive inductor-capacitor (LC) resonant circuit, of which the resonant frequency can be monitored wirelessly, leading to detections of gases in closed systems [8, 9].

Since this chapter is focused on sensors with the key structure of conductor-insulator-conductor, the discussion will not lay much emphasis on MOS capacitor sensors, which though will be mentioned in the text.

2. Fundamental Principles and Designs of Capacitive Sensors

Conventional capacitors, which are passive, two-terminal components, play indispensable roles in electrical circuits including storing electric charge, blocking direct current (dc) while allowing alternating current (ac) to pass, stabilizing output voltage, tuning frequency, etc. The essential part of a capacitor is a dielectric material (insulator) sandwiched between two electrodes that are made of good conductors. When a potential difference is built up between the two electrodes, equal but opposite charges are generated at the two electrode-insulator interfaces. The capacitance (C) of a capacitor is expressed as the amount of produced charge (Q) divided by the potential difference or voltage drop (V):

$$C = \frac{Q}{V} \tag{1}$$

Generally the insulator is homogeneous and the parallel-plate capacitor (Figure 1 (a)) is the most typical design in conventional capacitors (however in gas sensors neither of these holds well, relevant issues will be discussed in the next few sections). Therefore, the distribution of V along the electrode-to-electrode direction is linear, forming a constant electric field between the two parallel electrodes, which is equal to the two-dimensional charge density at the conductor-insulator interface (D_Q) divided by the absolute permittivity of the insulator (ε):

$$E = \frac{D_Q}{\varepsilon} \tag{2}$$

As commonly seen in textbooks, the distance between the two electrodes, or the thickness of the insulator, is defined as d and the area of the conductor-insulator interface is named A. Since $V=E\cdot d$, and $D_Q=Q/A$, the capacitance can then be characterized by the geometrical parameters of the capacitor (A and d) and the absolute permittivity of the dielectric medium:

$$C = \frac{\varepsilon \cdot A}{d} \tag{3}$$

To be used as a gas sensor, the capacitance must significantly and regularly change within a certain range of the analyte concentration in the environment. Thus, at least one of the three parameters in eqn. (3) (A, d, or ε) must be dependent on the gaseous substances of interest. Since A is always decides by the inflexible size of electrodes, changing the conductor-insulator interface area is uncommon in capacitive gas sensors. Most current capacitive gas sensors are based on the change of d or ε.

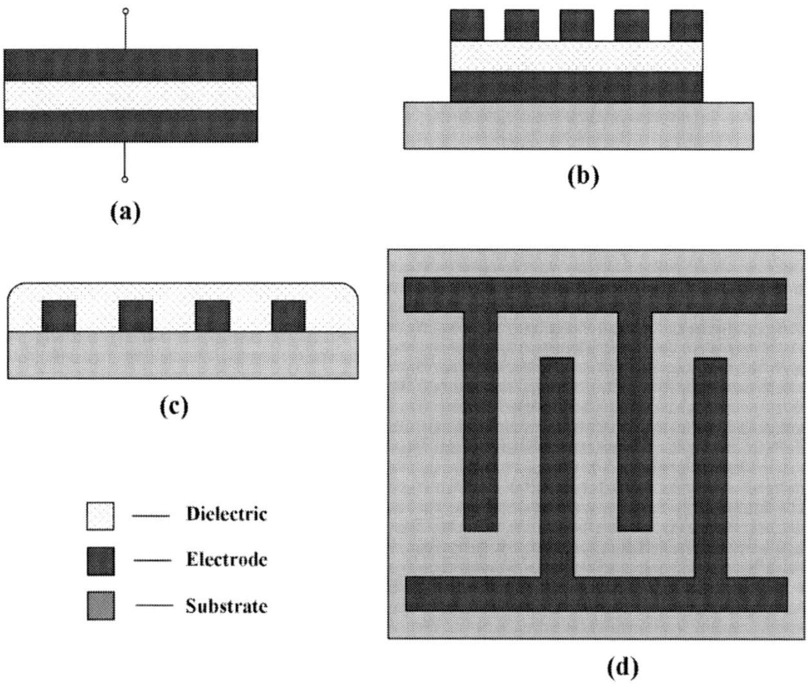

Figure 1. Fundamental structures of capacitors and capacitive sensors. (a) A simple planar capacitor. (b) A planar capacitor with holes throughout the top electrode, providing "windows" between the gas-sensitive dielectric material and the ambient. (c) The cross section of an interdigital capacitor (IDC) consisting of an interdigital electrode layer and a dielectric film (only four fingers are shown). (d) Top view of the interdigital design without the dielectric film.

Since these two parameters both depends on the insulator, the dielectric material in the capacitor should be sensitive to the analyte and functions as the sensing material. However, the parallel-plate capacitor in Figure 1 (a) is not always adopted in gas sensors and eqn. (3) works effectively only when A can be considered infinite compared to d. If the scale of d becomes close to that of A or even beyond, the electric field outside the electrode-electrode plane, or fringe capacitance, cannot be ignored.

In fact, few practical capacitive sensors are fabricated as the simple parallel-plate shape shown in Figure 1 (a), in which gas diffusion from ambient to the dielectric layer is rather blocked by the electrodes that are usually dense metals. To solve this blockade problem, alternative designs have been developed. In Figure 1 (b), holes throughout the top electrode or a porous top electrode is fabricated, allowing pathways for the analyte [10, 11, 12].

Moreover, a rigid substrate is always required for any device in order to provide support. The capacitor shown in Figure 1 (c), which can be considered as a two-layer structure from a design point of view, uses a pair of interdigital electrodes with a certain thickness and the dielectric material is deposited on this layout. Ignoring the capacitance at the corners, as well as fringing lines of the electric force, the approximate capacitance of such an interdigital capacitor (IDC) is:

$$C = \frac{\varepsilon \cdot z \cdot l \cdot (n-1)}{w} \tag{4}$$

where z is the thickness of the interdigital electrode layer, l is the overlap length of any two adjacent fingers, n is the total number of fingers on the overall layout, and w is the width of finger spacing. According to this equation, a small product of z, l and n leads to a decreased capacitance value. However, the IDC not only allows gases to sufficiently diffuse into the dielectric material, but has the advantage of reduced risk of short circuit due to mechanical stress.

More important, although eqn. (4) is a good estimate of the capacitance value of the IDC as long as z and l are fairly large. In cases that the film is thin (small z, e.g., sub-micron) but the interdigital distance is comparatively wide (large w), the electric field as well as the capacitance beyond the fringes of the electrode fingers could not be omitted, as mentioned before. With regards to fringing electric fields (FEF), computer simulations using finite-element method (FEM) have demonstrated the deviation of capacitance values of IDC from eqn. (4) [13].

Conversely, changes in fringe capacitance have been used to sense gases and this issue will be discussed in different cases in Section 6.

There are also several other types of capacitors, including coaxial cable, concentric spheres, etc., while their uses as sensors are rare.

3. Polymer-Based Capacitive Sensors and Humidity Detection

For the devices in which ε varies with the analyte, since the absolute permittivity is the product of dielectric constant (or relative permittivity) and the vacuum permittivity (8.854×10^{-12} $F\cdot m^{-1}$), the change of capacitance is actually the change of dielectric constant. Unlike resistive humidity sensors based on polyelectrolytes or conductive/semiconducting polymers [14], the dielectric material of the capacitive moisture sensors must be electrical insulating and insoluble in water, though able to reversibly absorb moisture in the environment. As a result, non-ionic but highly polar polymers that are hygroscopic, become good candidates.

Comparing to resistive polymer-based humidity sensors that are fabricated by dip-coatings, capacitors for water vapor detection usually assume increased manufacturing cost due to the casting or extrusion processes [15]. However, significant and almost linear responses of many capacitive sensors have made this type of devices more attractive than their resistive counterparts in humidity detection [15-17].

At room temperature, polymides, as a typical polymer group used in humidity sensors, have dielectric constants around 3 or slightly higher [18]. In contrast, the dielectric constant of pure water is large as 78.5 at 25°C [19]. This indicates that a little amount of water absorbed in the polymer dielectric would significantly increase the capacitance. As long as the equilibrium is achieved, the density of absorbed water in the polymer (C_W, in percentage terms, so it is less than unit) is proportional to the relative humidity (RH) in the ambient:

$$\frac{C_W}{RH} = D_W \tag{5}$$

D_W is the distribution coefficient of water between the dielectric material (polymer) and ambient (gas) phase. Since absorbed H_2O molecules should be distributed homogeneously in the medium, the dielectric constant of the polymer with water ($\varepsilon_{r,\,S}$) becomes the linear combination of the dielectric constant of the dry polymer ($\varepsilon_{r,\,P}$) and water ($\varepsilon_{r,\,W}$):

$$\varepsilon_{r,S} = \varepsilon_{r,P} \cdot (1 - C_W) + \varepsilon_{r,W} \cdot C_W \tag{6}$$

Combining eqn. (5) and (6) and assuming the volume expansion resulting from water absorption is trivial, the capacitance of the sensor is therefore linearly dependent on RH.

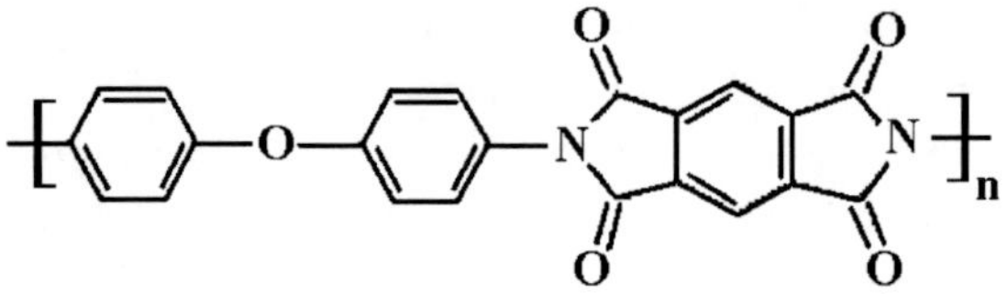

Figure 2. Kapton, a polyimide film developed by DuPont, with excellent stability in a broad range of temperature. It is being widely used in many electrical/mechanical areas, including humidity-sensing as well as supports for the sensors.

Due to the well-known capillary effect, under sufficient water vapor pressures, liquid-phase H_2O is inclined to condense at structures with very small radius, such as voids in polymers. Also, the very high hygroscopy of the sensing materials tends to irreversibly hold water [20, 21]. As a result, hysteresis becomes a serious problem in capacitive humidity sensors. Large water clusters inside the polymer not only reduce the performance but also cause deformation and shorten the lifetime of the device. The deformation and aging caused by water trapped in the dielectric materials can be solved by cross-linking, which moderately decreases the hygroscopy and enhances the performance of polymer-based resistive humidity sensors [22, 23, 24, 25]. The sensitivity was also found to be augmented by cross-linking, especially at high RH (> 80%), possibly due to the modified hygroscopy [25].

Being used as adhesives as well as photoresists in semiconductor industry [26], polyimides are the commonest group of sensing materials used for capacitive humidity sensors [12, 15-17, 20, 27-31]. The response of polyimide-based sensors is always linear and the detection limit is generally below 20% RH. Polymerized esters, such as cellulose acetate butyrate (CAB) [21, 32-34], polymethyl methacrylate (PMMA) [24, 25], polyvinyl crotonate [22], and polyethylene terephthalate (PETT) [35], are another group of dielectric materials frequently used in moisture-sensing capacitors, owing to their high polarity and non-ionic property. Previous studies have established humidity sensing models for both polyimides- and polymerized esters-based capacitive devices [15, 32]. Bedsides, capacitive humidity sensors based on polyethersulphone (PES) [15], polysulfone (PSF) [36], benzocyclobutene (BCB) [37, 38], hexamethyldisilazane (HMDS) [39], etc., have been reported.

State-of-the-art polymer-based capacitive humidity sensors are able to maintain linear responses down to 15% RH [15, 22, 28, 34, 35, 38] and hysteresis less than 2% has been found in some devices [15, 17]. Response time of a few seconds was reported for a 0% to 75% RH switch [31].

Besides humidity detection, a variety of polymers have been applied for capacitive sensing of volatile organic compounds such as alcohols, benzene derivatives, chemical warfare agents, etc [11, 40-42]. For chemicals with dielectric constant smaller than the sensing polymer, i. e., toluene absorbed in polyether urethane (PEUT) [40], the sensor capacitance decreased monotonically with the analyte concentration, instead of increasing. However, unlike their counterparts in humidity monitoring, non-linear response was frequently found in polymer-based capacitive sensors for organic vapor detections [11, 41, 42].

An acidic polycarbosilane with strong hydrogen bonding, or "HC" [43], which is able to filter out dimethyl methylphosphonate (DMMP, a simulant for the nerve agent sarin) in favor of other organic vapors, has been adopted as a superimposed layer to accumulate analytes, on single-wall carbon nanotube (SWCNT)-based capacitive gas sensors [44]. Several folds of increase of response were found on the HC-coated sensor, compared to the uncoated (this SWCNT-based capacitor-type sensor will be discussed in details in Section 6).

Although being successfully applied in capacitive humidity sensors, polymers are not the only type of materials in this kind of devices. Resistive sensors based on inorganic materials, especially metal oxides, is the dominant party in humidity detection [14]. Correspondingly, several moisture-sensitive capacitors using inorganic dielectric materials were reported during recent years. A capacitor with sol-gel processed γ-phase aluminum oxide (Al_2O_3) thick film as the dielectric/sensing material has been developed [45]. At the optimum measuring frequency, the capacitance was found to change by 38% as the moisture increased from 5 ppm to 95 ppm. The sensitivity was attributed to the nanoporous structure of the Al_2O_3 film. Similarly, a very recent study reported an Al_2O_3 capacitive humidity sensor with the detection limit down to 2.5 ppm [46]. However, these sensors were only characterized in trace humidity levels (< 1% RH). Capacitive humidity sensors have also been developed using silicon nanoporous pillar array [47] as well as silicon carbide (SiC) nanowires grown on the former [48] as sensing materials. The moisture sensitivity of these nano-scale porous materials can be ascribed to their large specific surface areas.

In addition, metal-organic frameworks (MOF) [49] have been applied in water vapor-sensitive capacitors. A 1 μm thick film of copper 1,3,5-benzenetricarboxylate ($Cu_3(BTC)_2$) crystals were grown on well-polished copper surface that served as one electrode and then aluminum was deposited on the top of the $Cu_3(BTC)_2$ film as another electrode, constructing a moisture-sensing capacitor [50]. The device demonstrated excellent linear response to water vapor. As the RH changed from 11% to 84%, the capacitance almost doubled. The hydrophilic part of the MOF structure was thought to be responsible for the good sensitivity to humidity and the depressed hysteresis is attributed to hydrophobic part of the $Cu_3(BTC)_2$ film, which may play a similar role to the cross-linking in polymer-based capacitive humidity sensors.

4. CO_2 AND SO_2 SENSING WITH AMINES

Chemical adsorption or chemisorption is a process in which chemical bonds are formed between the adsorbate and adsorbent. Many solid-state sensors utilize reversible chemisorption processes to detect specific substances and the dominant type is resistive devices. However, if a good adsorbent to a certain kind of gas molecule is almost completely non-conductive, it may not be appropriate to sever as the sensing material of a resistor-type device but possibly to be used in a capacitor- or mass-type sensor instead. As long as the change of dielectric constant of the dielectric material or adsorbent scales to the amount of adsorbate or analyte, the concentration of the analyte can be measured by the capacitance of the device. It should be noted that the phenomenon of "chemisorption" also occurs commonly in many capacitive sensors other than those discussed in this section, the CO_2 and SO_2 sensors are grouped here since they both contain amine groups in the sensing materials.

As an acidic compound, CO_2 is apt to react with amines. This indicates that materials with amine groups are good candidates for CO_2 chemisorption and the absorption of CO_2 by amines and has been used for CO_2 removal in space shuttles [51] and the atmosphere of earth (Carbon Sequestration) [52]. The initial application of this fundamental principle in gaseous CO_2 detection could be traced back to a mass sensor reported more than 20 years ago [53]. The first device using dielectric materials with high density of adjacent primary amine groups (R-NH_2) on the surface was reported in 1995 [54]. At the temperature range of 60-80°C, the capacitance of an IDC using a mixture of 3-aminopropyltrimethoxysilane (AMMO) and propyltrimethoxysilane (PTMS) as the dielectric material, decreased with CO_2 and the relative change of capacitance in response to 1% CO_2 was about 2% at 70°C. It was proposed that AMMO interacts reversibly to CO_2 while PTMS reduces the cross-sensitivity to H_2O. Although the reaction between the basic R-NH_2 and acidic CO_2 or hydrated CO_2 (carbonic acid) is regarded spontaneous in theoretical chemistry, detailed sensing mechanisms were expected.

Published almost at the same time as the novel device discussed above, Leal et al. found that a ring-shaped ionic structure could be formed between two adjacent primary amine groups rooted on the same substrate, being bridged by a CO_2 molecule [55]. As shown in Figure 3, after reacting with CO_2, one amine group becomes an ammonium ion while another changes into a carbamic acid and a cyclic carbamate is formed. Continuing reacting with CO_2 and H_2O, the adjacent primary amine pair finally converts to ammonium bicarbonates. At moderately elevated temperatures (e.g., 40-140°C), the overall process is completely reversible. As a result, this type of CO_2 sensors reported in 1990s operated at above room temperature and an on-chip heater was included as part of the device [54, 56]. Nonetheless, recent investigations have suggested lower operating temperatures. Increasing the length of the alkyl chain connected to the amine ("R" in Figure 3) could effectively decrease the working temperature down to 25-35°C [57]. Moreover, a room-temperature reversible formation of the cyclic carbamate in the first stage of amine-CO_2 interaction in Figure 3 was proposed and believed to be responsible for the rather stabilized performance after the 7:3 (volume ratio) AMO-PTMS layer was stored in humid air for seven months following manufacturing [58].

According to the mechanism in Figure 3, materials with high densities of continuously distributed primary amine groups are also subject to water vapor adsorption and the cross-sensitivity to humidity is a major problem. Furthermore, the CO_2 signal magnitude is appreciably attenuated with storage time as the adsorbed moisture accumulates [58]. The hydrophobic PTMS, as mention above, has been mixed with the sensing material (AMMO) to suppress the interference of moisture, which, however, could not be reduced to insignificant and a separate humidity sensor was previously used for numerical compensation [56]. Besides of lowering the operating temperature, increased length of the alkyl chain was found to be able to increase CO_2 sensitivity while relatively reduce H_2O sensitivity [57].

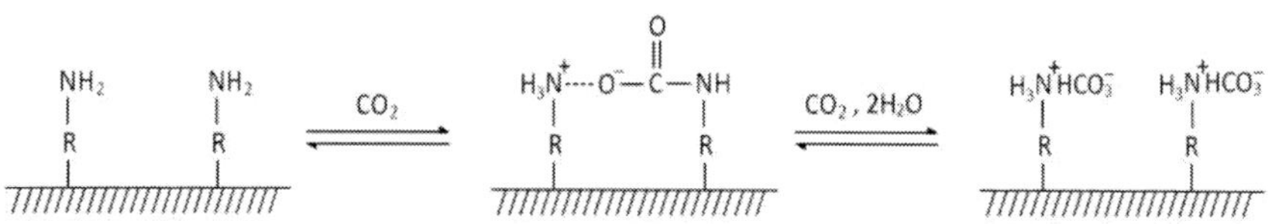

Figure 3. Two-stage chemisorption of CO_2 on a surface with adjacent primary amine groups.

The room-temperature reversible formation of carbamate suggested in Ref. [58] was also thought to somewhat improve the stability. As reported by another recent study [59], 7-8 μm-thick A3EO (3-aminopropyltriethoxysilane) demonstrated higher sensitivity and more stable performance at the presence of moisture than the AMMO-PTMS mixture with close thickness. However, the interference of mid-to-high RH (>20%) was still pronounced.

Besides causing changes in the dielectric constant, the chemisorption of CO_2 can alter the optical properties of materials with multiple primary amine groups and the CO_2 concentration in the ambient can be measured by optical approaches. Instead of using silane derivatives, Rocchia et al. adopted porous silicon films modified with 3-aminopropanol as the dielectric and the amine-terminated silicon surface demonstrated measurable optical responses to CO_2 [60]. Moreover, Star et al. fabricated a carbon nanotube (CNT)-based CO_2 sensor with polyethylenimine (PEI) containing 25% primary amine groups as the sensing material [61]. The overall device was a field effect transistor (FET) using CNT as the conduction channel and the polymer was coated on the CNT. Under fixed gate and source-drain voltages, the current flow between the drain and source decreased monotonically with CO_2 concentration in air at 80% RH. Most possibly, the sensing reaction in Figure 3 deformed the polymer layer and the electron doping of the CNT beneath is consequently changed. Hygroscopic starch was used along with PEI in order to enhance the CO_2-H_2O-amine reactions.

A room-temperature CO_2-sensitive capacitor with Teflon AF 2400^{TM} (a product of DuPont) as the dielectric material is reported [62]. The detection limit showed down to 420 ppm CO_2 while the Teflon AF 2400^{TM} was also very sensitive to water vapor, with the response over 30 times than that to CO_2. However, detailed mechanism was not described and it was undetermined if any significant reversible chemisorption (like the interaction between R-NH_2 and CO_2) was involved in the sensing process.

Carbon dioxide is apparently not the only kind of gas analyzed by amine-type capacitors. Back to 1940s, it was discovered that the reversible reaction between trimethylamine ($N(CH_3)_3$) and SO_2 forms a complex at room temperature [63] and the complex was characterized rather thoroughly in the 1960s [64]. Even before the emergence of CO_2 sensors based on primary amine dielectric materials, the capacitance of tertiary amine (NR_3) was found to be sensitive to sulfur dioxide (SO_2) [65] and the sensing mechanism has been attributed the formation of the amine-SO_2 complex [66]. Different from the CO_2 sensing reaction in Figure 3, which reduces the relative permittivity of R-NH_2 materials, the capacitance of this type of sensor increases with SO_2

concentration. A temperature-dependence model was also reported [67]. More recent studies have shown that the NR_3-SO_2 interaction can also be adopted into QCM sensors for sulfur dioxide [68, 69].

5. GAS-SENSING CAPACITORS USING P-N JUNCTIONS

In the early 1990s, it was found that capacitors based on some binary mixtures of metal oxides were considerably sensitive to CO_2 [70, 71] or nitrogen oxides (NO_X) [72], though the response of either component alone showed insignificant. This type of devices usually operates at elevated temperatures (100-1000°C) and the oxides can be regarded as high-temperature semiconductors. The change of sensitivity with temperature has been proved to be bell-shaped; the sensitivity rises first and then starts to drop at a maximum value, at which the temperature could be regarded as optimum. The optimum operating temperature changes from one mixture to another. Among all the tested mixtures, the highest response to NO_X was found on capacitors based on NiO-ZnO [72] and CuO-$BaTiO_3$ is quite sensitive to CO_2 and the most commonly used [70, 73-75], although other Perovskite compounds [76] such as $SrTiO_3$, $BaSnO_3$, or non-Perovskites have also been tested as substitutes to CuO-$BaTiO_3$ and CuO-$BaSnO_3$ demonstrated an even higher sensitivity [77].

Unlike the amine-type sensors discussed in Section 4, this type of devices demonstrated rather low sensitivity to water vapor, in the case of either CO_2 [70] or NO_X sensing [72]. With an approximately 1 mol% of silver additive, the response to 2% CO_2 of a capacitor based on CuO-$BaTiO_3$ could be amplified to several tens of times and the relative sensitivity to H_2O became very small [73]. Capacitors based on triple mixtures including CuO, $BaTiO_3$, and CeO_2 have also been developed and interference from 0% to 60% RH at 30°C was negligible [74]. The introduction of silver or CeO_2 could effectively prevent degradation, which is a noticeable problem in this type of devices.

The key principle can be described in Figure 3. A homogeneous mixture of a p-type and an n-type oxide is sandwiched. Since the oxides are semiconductors, the capacitance of the device is definitely not measured by the absolute distance between the two electrodes. At the contact of a p- and an n-type particle, a p-n junction with unilateral conductivity is formed. Due to the high density and even distribution of the p-n contacts, there should be another p-n junction with a reverse orientation in the inter-electrode conduction path and therefore the whole mixture become non-conductive. The insulator or dielectric of this kind of devices is actually the sum of the p-n junctions along the direction perpendicular to the electrode plane. The adsorption of analytes (e.g., CO_2 or NO_X) reduces (sometimes enlarges) the thickness of p-n junctions and therefore increases (or decreases) the capacitance.

Although in Figure 4, the device response to the analyte is shown as an enhancement of the capacitance, actual sensing behaviors depend on the material compositions as well as the analyte. In a switch of pure air to 2% CO_2, a capacitance increment around three times was observed on an equimolar (1:1) mixture of CuO-$BaTiO_3$ at the optimum operating temperature, but a 1:3 blend of Co_3O_4-$BaTiO_3$ caused about a three folds-decrease of the capacitance at its own optimum operating temperature [70], indicating a broadening of the p-n junctions. Also, when 100 ppm NO was introduced into pure air, the capacitance of a device based on NiO-$BaTiO_3$ only increased by less than a half, whereas a NiO-ZnO mixture could attain a capacitance reduction of twenty times [72].

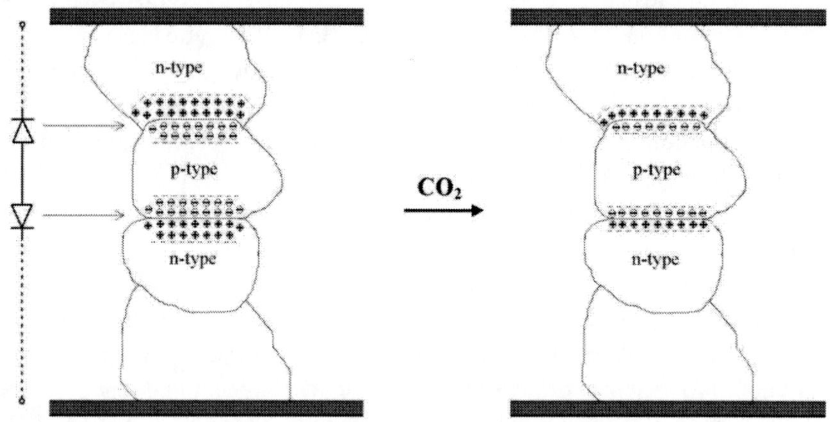

Figure 4. The fundamental principle of CO_2 sensing by a device composing of p-n junctions widely distributed in the materials between the two electrodes. Since the p-n junctions are randomly dispersed, direct current (DC) cannot flow either way between the two electrodes and therefore the device can be regarded as a capacitor. In order to simplify the discussion, assuming two back-to-back p-n junctions located at the interfaces between p-type (e.g., CuO) and n-type (e.g., $BaTiO_3$) particles are approximately parallel to the electrode plane. The junction areas, or the space charge regions (positive charges in the n-side while negative charges in p-side), act as the insulator between the two electrodes and the sum of the junction widths (inside the dash lines) is the insulator thickness. As a result, if the switch from air to CO_2 reduces the junction width, the capacitance of the device is augmented.

The detailed mechanism on the molecular level remains unclear. For CO_2 sensing, reversible carbonation of CuO-$BaTiO_3$ has been proposed as the dominant chemical reaction in the sensing process and supported by the bell-shaped sensitivity-temperature curves as well as the effects resulting from the silver additive [71, 73], while a later study suggested a possible surface adsorption, since no significant evidence of carbonating valency states could be shown in the X-ray photospectrometry (XPS) for the sensor sample [75]. Nevertheless, the fundamental principle of this kind of devices as shown in Figure 4, in which the contact between the p- and n-type materials plays the key role, is supported by almost all published studies.

From the discussions above, one may notice that the device in Figure 4 operates rather similarly to the MOS capacitor-type gas sensor (mostly for H_2 sensing) [7, 78-80], of which the capacitance is altered by the thickness of the depletion region of the semiconductor beneath the insulator, which depends on the bias voltage as well as hydrogen adsorbed at the metal-insulator interface [81].

In contrast, the depletion region in MOS devices, which are usually fabricated on a single-crystal semiconductors, is uniform; while in the devices discussed in this section, the p-n junctions are distributed randomly due to the intrinsic structure of films composed of a large number of particles. Also, a catalytic metal gate is required in the MOS sensors to induce the sensitivity and a certain bias voltage must be presented for measurements. The modulation of space charge region by adsorbed species will be discussed again in a carbon nanotube-based gas sensor in Section 6.

Besides, the fundamental sensing principle of p-n junction-based capacitors is somewhat alike to that of varistor-type gas sensors, in which the height of the double Schottky barrier at the grain boundaries could be changed by adsorptions of gas molecules [82]. In fact, both the impedance and resistance of the gas-sensing capacitor in Figure 4 drops, in the meantime of the capacitance increase. However, a varistor uses I-V measurements to sense gases, which requires very high voltages (usually over 100 V).

The approach of mixing $BaTiO_3$ with various semiconducting oxides has also been used to develop resistor-type CO_2 sensors, in which the device resistance increases with CO_2 concentration [83, 84]. Selective detections of gases other than CO_2 or NO_X using devices based on the change at the interface of p- and n-type materials have also been reported. In a resistive sensor, the reduction of p-n junctions formed at CuO-SnO_2 contacts was thought to be responsible for the ultra-high sensitivity (up to tens of thousands of folds) to H_2S [85].

6. Novel Capacitive Sensors Based on Carbon Nanotubes

Carbon nanotube (CNT) is one of the commonest and most attractive nano-sized structures synthesized by human kind, with controlled dimensions and tunable physical properties [86]. The molecular structure of a single-wall CNT (SWCNT) can be simply regarded as a graphene sheet rolled into a seamless cylinder with a diameter of nanometers or even below. According to the similarity in the bondings, CNT can be regarded as a polymorph of graphene, fullerene, graphite, etc. The top view of a multi-wall CNT (MWCNT) is like a helix, since the graphene sheet rolls itself into multiple layers. The MWCNT was discovered in 1991 by Iijima [87]. Two years later, the synthesis of SWCNT was reported by Iijima et al. and Bethune et al., independently [88, 89]. In 1995, Smalley group functionalized MWCNT into a field emitter and used the tube as the tip in scanning probe microscopy in the next year, revealing its potential in electronic devices [90, 91]. During the past 15 years, nanotube-based field effect transistors (NTFET), in which single semiconducting SWCNT [92, 93] or modified CNT bundles [94] serve as the source-drain channel, were successfully fabricated. High-performance CNT thin-film transistors and integrated circuits have also been developed [95].

Besides single or multiple rolling layers, two fundamental types of CNT are discerned based on the structure-dependent electronic properties: metallic and semiconducting [96]. Owing to its tunable conducting properties in response to electric field or adsorption of special chemicals, semiconducting CNT, which could also be doped in order to adjust the electronic characteristics [97], are preferred in most CNT-based microelectronic devices, including gas sensors.

At the beginning of the 21th century, resistance of SWCNT bundles, films, or mats were found to change with the concentrations of gaseous analytes including ammonia (NH_3) [98], NO_2 [98], O_2 [99, 100], and H_2O [101]. Gas sensing properties of CNT were thoroughly investigated in these studies. Kong et al. found that the conductance of individual SWCNT increased with NO_2 while decreased with NH_3 and the nanotube was confirmed to be p-type [98]. While as-prepared nanotubes were frequently reported to be strongly p-type, Collins et al. noted that it may not be intrinsic but a consequence of oxygen doping in the air [99]. CO_2 sensors based on NTFET have also been developed [61], as we have discussed in Section 3, although CNT was not used as the sensing material.

Varghese et al. presented an early trial of using CNT to interact directly with the analyte in a capacitive microelectronic device [102]. The basic design of the device was a pair of interdigital metal electrodes fabricated on a conventional printed circuit board. A silicon oxide (SiO_2) layer was deposited on the electrodes to ensure insulation and then a 0.2 mm-thick SiO_2 layer with MWCNTs dispersed randomly inside (weight ratio of MWCNT and SiO_2 was 2:3), was superimposed on the former. As a result, only fringing electric fields (FEF) existed in the MWCNT-SiO_2 layer. According to the Cole-Cole impedance plot, the overall device was equivalent to a parallel resistor-capacitor circuit. MWCNTs used in the layer were believed to be a mixture of metallic and p-type semiconducting nanotubes. The resistance increased with NH_3, H_2O, and CO_2, indicating the adsorptions of these electron donors to the MWCNTs reduce the conductivity of the p-type nanotubes. For the capacitance change, the widths of the Schottky junctions formed at the metallic CNT-p-type CNT contacts played an importance role. These junctions contributed significantly to the total device capacitance and could be modulated by the adsorbed gas molecules, as that occurs in the p-n junction- or MOS-based devices in Section 5. Since NH_3, H_2O, and CO_2 down-regulated the junction widths, the capacitance value increased with these electron donor-type gases.

A later study of the same research group found that the conductivity of MWCNT-SiO_2 layer increased slightly with O_2, which tends to deprive electrons from the semiconducting MWCNTs, confirming the p-type behaviors [9]. In this study, the sensor unit was passive and an inductor was included along with the MWCNT-SiO_2 device in order to form an LC circuit. The sensing signal was monitored remotely, in the form of the variation of the resonant frequency of the LC circuit, through an outside loop antenna, as shortly discussed at the end of Section 1. Wireless sensing approach was also reported in a CNT-coated electromagnetic resonator sensor, which transduced the adsorption of NH_3 on the nanotubes into frequency shift [103].

In 2005, Snow et al. reported a novel capacitive device that was able to sense multiple organic vapors with quite fast responses at trace levels (e.g., ppb DMMP) [44]. Besides the always positive capacitance change to many gases at very low concentrations, the sensor also showed swift recoveries upon the removals of analytes (e.g., 90% recovery time for 50 ppb DMMP was less than 4 seconds), which can be regarded as a significant breakthrough, since CNT-based gas sensors developed in the early few years of this century always suffered slow recoveries at room temperature [98, 99, 104] and light had to be used to stimulate gas desorption from CNTs [105]. The device used a degenerately doped silicon (this means its conductivity is comparable to metals) as the bottom electrode as well as the substrate. Through thermal oxidation, a 250 nm-thick SiO_2 was fabricated on the silicon and then an SWCNT network was grown on the former using chemical vapor deposition (CVD). Finally, an interdigital palladium (Pd) electrode pair was deposited on the SWCNT. As discussed in Section 3, HC was coated on top of the Pd electrodes to accumulate analytes in some measurements.

It should be noted that the top electrode of the capacitor was not the pair of Pd electrodes but the SWCNT network beneath. An essential advantage of this sensor architecture is that it provides the ability to perform conductance and capacitance measurements simultaneously for each chemical vapor of interest. In addition, unlike all the conductor-insulator-conductor capacitive sensors discussed previously in this chapter, the sensing material (CNT) of this device was not the dielectric but one of the electrodes. This is somewhat similar to MOS sensors that use catalytic metal (mostly Pd) as the top electrode of the gate capacitor [7, 80].

Field-induced polarization of surface dipoles is believed to be responsible for the capacitance change. Vapor molecules adsorb at the SWCNT surface, where the FEF is the strongest, are polarized by the intense FEF and cause increases in the device capacitance. This mechanism has been proved by the exponential correlation between the relative capacitance change and the value of vapor concentration divided by the equilibrium vapor pressure, as observed in quite a few analytes tested within a rather wide concentration range [44, 104].

Charge transfer from the adsorbed analytes to the SWCNT does affect the capacitance, while the resulted effects on the device response only accounts for around 10% of the total signal. In a later report, conductance measurements through the interdigital electrode pair, though demonstrated sensitivity to analytes, were found to be notably unreliable due to the charge effects, and therefore the capacitance-based sensing should be much more practical for this type of devices [106].

Snow group's another work using the same device structure but titanium/gold interdigital electrodes not only confirmed the field-induced polarization mechanism but also found that the device response, as well as the sensitivity to analytes, were dominated by the gas molecule adsorption at the defect sites in the nanotubes, which could be controlled in order to enhance the sensitivity [107].

CONCLUSION

We have reviewed multiple types of gas sensing capacitors with the structure of conductor-insulator-conductor. Besides materials and devices, the interactions between sensing materials and analytes are discussed in details. Application issues, such as the interferences, operating temperatures, costs, etc., are evaluated as well. Discussions also incorporate non-capacitive sensors including resistors, MOS devices, and other types, which possess identical or similar sensing mechanisms to the reviewed capacitors.

REFERENCES

[1] Comini, E; Faglia, G; Sberveglieri G.; Ed.; Solid State Gas Sensing; Springer Science+Business Media LLC: New York, NY, 2009.
[2] Janata, J. Principles of Chemical Sensors; Springer Science+Business Media LLC: New York, NY, 2009.
[3] Moseley, P. T. *Meas. Sci. Technol.* 1997, 8, 223-237.
[4] Lucklum, R.; Hauptmann, P. *Sens. Actuators*. B 2000, 70, 30-36.
[5] Weppner, W. *Sens. Actuators.* 1987, 12, 107-119.
[6] Huang, X. -J.; Choi, Y. -K. *Sens. Actuators*. B 2007, 122, 659-671.
[7] Lundstrom, I.; Sundgren, H.; Winquist, F.; Eriksson, M.; Krantz-Rulcker, C.; Lloyd-Spetz, A. *Sens. Actuators*. B 2007, 30, 247-262.
[8] Ong, K. G.; Grimes, C. A.; Robbins, C. L.; Singh R. S. *Sens. Actuators*. A 2001, 93, 33-43.
[9] Ong, K.; Zeng, K.; Grimes, C. *IEEE Sens. J.* 2002, 2, 82-88.
[10] Delapierre, G.; Grange, H.; Chambaz, B.; Destannes, L. *Sens. Actuators*. 1983, 4, 97-104.
[11] Patel, S. V.; Mlsna, T. E.; Fruhberger, B.; Klaassen, E.; Cemalovic, S.; Baselt, D. R. *Sens. Actuators*. B 2003, 96, 541-553.
[12] Lazarus, N.; Fedder, G. K. *J. Micromech. Microeng.* 2011, 21, 065028/1-065028/9.
[13] Lin, J.; Moller S.; Obermeier, E. *Sens. Actuators*. B 1991, 5, 223-226.
[14] Chen, Z.; Lu C. *Sens. Lett.* 2005, 3, 274-295.
[15] Harrey, P. M.; Ramsey, B. J.; Evans, P. S. A.; Harrison, D. J. *Sens. Actuators*. B 2002, 87, 226-232.
[16] Ralston, A. R. K.; Klein, C. F.; Thomas, P. E.; Denton, D. D. *Sens. Actuators*. B 1996, 34, 343-348.

[17] Dokmeci, M; Najafi, K. *J. Microelectromech.* S. 2001, 10, 197-204.
[18] Maier, G. *Prog. Polym. Sci.* 2001, 26, 3-65.
[19] Uematsu, M.; Frank, E. U. *J. Phys. Chem. Ref. Data.* 9, 1980, 1291-1307.
[20] Sakai, Y.; Sadaoka, Y.; Matsuguchi, M. *Sens. Actuators.* B 1996, 35, 85-90.
[21] Matsuguchi, M.; Umeda, S.; Sadaoka, Y.; Sakai, Y. *Sens. Actuators.* B 1998, 49, 179-185.
[22] Matsuguchi, M.; Sadaoka, Y.; Nuwa, Y.; Shinmoto, M.; Sakai, Y.; Kuroiwa, T. *J. Electrochem. Soc.* 1994, 141, 614-618.
[23] Matsuguchi, M.; Shinmoto, M.; Sadaoka, Y.; Kuroiw, T.; Sakai, Y. *Sens. Actuators.* B 1996, 34, 349-355.
[24] Matsuguchi, M.; Kubo, S.; Sakai, Y. *Electrochemistry.* 1999, 67, 170-174.
[25] Roman, C.; Bodea, O.; Prodan, N.; Levi, A.; Cordosa, E.; Manoviciu, I. *Sens. Actuators.* B 1995, 25, 710-713.
[26] Ghosh, M.; Mittal K. L.; Ed.; Polyimides: Fundamentals and Applications; Marcel Dekker Inc.: New York, NY, 1996.
[27] Matsuguchi, M.; Sadaoka,Y.; Nosaka, K.; Ishibashi, M.; Sakai, Y.; Kuroiwa, T.; Ito, A. *J. Electrochem. Soc.* 1993, 140, 825-829.
[28] Liu, X.; Eriksent, G. F.; Leistikot, O. *J. Micromech. Microeng.* 1995, 5, 147-149.
[29] Ingram, J. M.; Greb, M.; Nicholson, J. A.; Fountain, A. W. *Sens. Actuators.* B 2003, 96, 283-289.
[30] Lee, C. -Y.; Wu, G. -W.; Hsieh, W. -J. *Sens. Actuators.* A 2008, 147, 173-176.
[31] Lee, H.; Lee, S.; Jung, S.; Lee, J.; *Sens. Actuators.* B 2011, 154, 2-8.
[32] Grange, H.; Beith, C.; Boucher, H.; Delapierre, G. *Sens. Actuators.* 1987,12, 291-296.
[33] Sadaoka, Y.; Matsugucgi, M.; Sakai, Y.; Takahashi, K. *J. Mater. Sci. Lett.* 1988, 7, 121-124.
[34] Oprea, A.; Barsan, N.; Weimar, U.; Bauersfeld, M. -L.; Ebling, D.; Wollenstein, J. *Sens. Actuators.* B 2008, 132, 404-410.
[35] Perez, J. M. M.; Freyre, C. Sens. Actuators B 1997, 42, 27-30.
[36] Kuroiwa, T.; Miyagishi, T.; Ito, A.; Matsuguchi, M.; Sadaokab, Y.; Sakai, Y. *Sens. Actuators.* B 1995, 25, 692-695.
[37] Te'telin, A.; Pellet, C.; Laville, C.; N'Kaoua, G. *Sens. Actuators.* B 2003, 91, 211-218.

[38] Zampetti, E.; Pantalei, S.; Pecora, A.; Valletta, A.; Maiolo, L.; Minotti, A.; Macagnano, A.; Fortunato, G.; Bearzotti, *Sens. Actuators*. B 2009, 143, 302-307.
[39] Kraus, F.; Cruz, S.; Muller, J. *Sens. Actuators*. B 2003, 88, 300-311.
[40] Hierlemann, A.; Lange, D.; Hagleitner, C.; Kerness, N.; Koll, A.; Brand, O.; Baltes, H. *Sens. Actuators*. B 2000, 70, 2-11.
[41] Mlsna, T. E.; Cemalovic, S.; Warburton, M.; Hobson, S. T.; Mlsna, D. A.; Patel, S. V. *Sens. Actuators*. B 2006, 116, 192-201.
[42] Kitsara, M.; Goustouridis, D.; Chatzandroulis, S.; Chatzichristidi, M.; Raptis, I.; Ganetsos, T.; Igreja, R.; Dias, C. J. *Sens. Actuators*. B 2007, 127, 186-192.
[43] Novak, J. P.; Snow, E. S.; Houser, E. J.; Park, D.; Stepnowski J. L.; McGill R. A. *Appl. Phys. Lett.* 2003, 83, 4026-4028.
[44] Snow, E.; Perkins, F.; Houser, E.; Badescu, S.; Reinecke, T. *Science.* 2005, 307, 1942-1945.
[45] Mistry, K. K.; Saha, D.; Sengupta, K. *Sens. Actuators*. B 2005, 106, 258-262.
[46] Islam, T.; Kumar, L.; Khan, S. A. *Sens. Actuators*. B 2012, 173, 377-384.
[47] Li, L. Y.; Dong, Y. F.; Jiang, W. F.; Ji, H. F.; Li, X. J. *Thin Solid Films.* 2008, 517, 948-951.
[48] Wang, H. Y.; Wang, Y. Q.; Hu, Q. F.; Li, X. J. Capacitive humidity sensing properties of SiC nanowires grown on silicon nanoporous pillar array, *Sens. Actuators*. B 2012, 166-167, 451-456.
[49] James, S. L. *Chem. Soc. Rev.* 2003, 32, 276-288.
[50] Liu, J.; Sun, F.; Zhang, F.; Wang, Z.; Zhang, R.; Wang, C.; Qiu, S. *J. Mater. Chem.* 2011, 21, 3775-3778.
[51] Satyapal, S.; Filburn, T.; Trela, J.; Strange, J. *Energ. Fuel.* 2001, 15, 250-255.
[52] Gray, M. L.; Soong, Y.; Champagne, K. J.; Pennline, H. W.; Baltrus, J. *Int. J. Environ. Technol. Manage.* 2004, 4, 82-88.
[53] Fatibello-Filho, O.; de Andrade, J. F.; Suleiman, A. A.; Guilbault, G. G. *Anal. Chem.* 1989, 61, 746-748.
[54] Mutschall, D. Obermeier, E. *Sens. Actuators*. B 1995, 25, 412-414.
[55] Leal, O.; Bolivar, C.; Ovalles, C.; Garcia, J. J.; Espidel, Y. *Inorg. Chim. Acta.* 1995, 240, 183-189.
[56] Endres, H. -E.; Hartinger, R.; Schwaiger, M.; Gmelch, G.; Roth, M. *Sens. Actuators*. B 1999, 57, 83-87.

[57] Patel, S. V.; Hobson, S. T.; Cemalovic S.; Mlsna, T. E. *J. Sol-Gel. Sci. Technol.* 2010, 53, 673-679.
[58] Stegmeier, S.; Fleischer, M.; Tawil, A.; Hauptmann, P.; Egly, K.; Rose, K. *Sens. Actuators.* B 2011, 154, 270-276.
[59] Stegmeier, S.; Fleischer, M.; Tawil, A.; Hauptmann, P. *Sens. Actuators.* B 2010, 148, 450-458.
[60] Rocchia, M.; Garrone, E.; Geobaldo, F.; Boarino, L.; Sailor, M. *J. Phys. Stat. Sol.* A 2003, 197, 365- 369.
[61] Star, A.; Han, T. -R.; Joshi, V.; Gabriel, J. -C. P.; Grüner, G. *Adv. Mater.* 2004, 16, 2049-2052.
[62] Kebabian P. L.; Freedman, A. *Meas. Sci. Technol.* 2006, 17, 703-710.
[63] Burg, A. B. *J. Am. Chem. Soc.* 1943, 65, 1629-1635.
[64] *Christian S. D.; Grundnes, J. Nature. 1967, 214, 1111-1112.*
[65] Lin, J.; Moller S.; Obermeier, E. *Sens. Actuators.* B 1991, 5, 219-221.
[66] Endres, H. -E.; Drost, S.; Hutter, F. *Sens. Actuators.* B 1994, 22, 7-11.
[67] Horn, M.; Pichlmaier, J.; Trankler, H. -R. *Sens. Actuators.* B 1995, 25, 400-402.
[68] Matsuguchi, M.; Sakurada, K.; Sakai, Y. *J. Appl. Polym. Sci.* 2003, 88, 2982-2987.
[69] *Dultsev, F. N.; Sveshnikova, L. L. Sens. Actuators.* B 2007, *120, 434-438.*
[70] Ishihara, T.; Kometani, K.; Mizuhara, Y.; Takita, Y. *Sens. Actuators.* B 1991, 5, 97-102.
[71] Ishihara, T.; Kometani, K.; Hashida, M.; Takita, Y. *J. Electrochem. Soc.* 1991, 138, 173-176.
[72] Ishihara, T.; Sato, S.; Takita, Y. *Sens. Actruators.* B 1995, 25, 392-395.
[73] Ishihara, T.; Kometani, K.; Nishi, Y.; Takita, Y. *Sens. Actruators.* B 1995, 28, 49-54.
[74] Matsubara, S.; Kaneko, S.; Morimoto, S.; Shimizu, S.; Ishihara, T.; Takita, Y. *Sens. Actruators.* B 2000, 65, 128-132.
[75] Liao, B.; Wei, Q.; Wang, K.; Liu, Y. *Sens. Actruators.* B 2001, 80, 208-214.
[76] Navrotsky, A. *Chem. Mater.* 1998, 10, 2787-2793.
[77] Ishihara, T.; Kometani, K.; Mizuhara Y.; Takita, Y. *J. Electrochem. Soc.* 1992, 139, 2881-2885.
[78] Lundstrom, I. *Sens. Actuators.* 1981, 1, 403-426.
[79] Lundstrom, I.; Svensson, C.; Spetz, A.; Sundgren H.; Winquist, F. *Sens. Actruators.* B 1993, 13, 16-23.

[80] Lu C.; Yu, P. in Gas Sensors: Developments, Efficacy and Safety; Qiu X.; Ed.; Nova Science Publishers: Hauppauge, NY, 2011; pp.179-202.

[81] Ekedahl, L. -G.; Eriksson, M.; Lundstrom, I. *Acc. Chem. Res.* 1998, 31, 249-256.

[82] Egashira, M.; Shimizu, Y.; Takao, Y.; Sako, S. *Sens. Actuators.* B 1996, 35, 62-67.

[83] Haeusler, A.; Meyer, J. -U. *Sens. Actuators.* B 1996, 34, 388-395.

[84] Keller, P.; Ferkel, H.; Zweiacker, K.; Naser, J.; Meyer, J. -U.; Riehemann, W. *Sens. Actuators.* B 1999, 57, 39-46.

[85] Tamaki, J.; Maekawa, T.; Miura, N.; Yamazoe, N. *Sens. Actuators.* B 1992, 9, 197-203.

[86] Jorio, A.; Dresselhaus, G.; Dresselhaus, M. S.; Ed.; Carbon Nanotubes: Advanced Topics in the Synthesis, Structure, Properties and Applications; Springer: New York, NY, 2008.

[87] Iijima, S. *Nature.* 1991, 354, 56-58.

[88] Iijima, S.; Ichihashi, T. *Nature.* 1993, 363, 603-605.

[89] Bethune, D. S.; Klang, H.; de Vries, M. S.; Gorman, G.; Savoy, R.; Vazquez J.; Beyers, R. *Nature.* 1993, 363, 605-607.

[90] Rinzler, A. G.; Hafner, J. H.; Nikolaev, P.; Lou, L.; Kim, S. G.; Tomanek, D.; Nordlander, P.; Colbert, D. T.; Smalley, R. E. *Science.* 1995, 269, 1550-1553.

[91] Dai, H. J.; Hafner, J. H.; Rinzler, A. G.; Colbert, D. T.; Smalley, R. E. *Nature.* 1996, 384, 147-150.

[92] Tans, S. J.; Verschueren, A. R. M.; Dekker, C. *Nature.* 1998, 393, 49-52.

[93] Chen, Z.; Appenzeller, J.; Knoch, J.; Lin, Y. -M.; Avouris, P. *Nano Lett.* 2005, 5, 1497-1502.

[94] Collins, P. G.; Arnold, M. S.; Avouris, P. *Science.* 2001, 292, 706-709.

[95] Sun, D. -M.; Timmermans, M. Y.; Tian, Y.; Nasibulin, A. G.; Kauppinen, E. I.; Kishimoto, S.; Mizutani T.; Ohno Y. *Nat. Nanotechnol.* 2011, 6, 156-161.

[96] Wilder, J. W. G.; Venema, L. C.; Rinzler, A. G.; Smalley R. E., Dekker, C. *Nature.* 1998, 391, 59-62.

[97] Peng S.; Cho K. *Nano Lett.* 2003, 3, 513-517.

[98] Kong, J.; Franklin, N. R.; Zhou, C.; Chapline, M. G.; Peng, S.; Cho, K.; Dai, H. J. *Science.* 2000, 287, 622-625.

[99] Collins, P. G.; Bradley, K.; Ishigami, M.; Zettl, A. *Science.* 2000, 287, 1801-1804.

[100] Sumanasekera, G. U.; Adu, C. K. W.; Fang, F.; Eklund P. C. *Phys. Rev. Lett.* 2000, 85, 1096-1099.

[101] Zahab, A.; Spina, L.; Poncharal, P. Phys. Rev. B 2000, 62, 10000-10003.
[102] Varghese, O.; Kichambre, P.; Gong, D.; Ong, K.; Dickey, E.; Grimes, C. *Sens. Actuators*. B 2001, 81, 32-41.
[103] Chopra, S.; Pham, A.; Gaillard, J.; Parker, A.; Rao A. M., *Appl. Phys. Lett.* 2002, 80, 4632-4634.
[104] Snow, E.; Perkins, F.; Robinson, J. *Chem. Soc. Rev.* 2006, 35, 790-798.
[105] Chen, R. J.; Franklin, N. R.; Kong, J.; Cao, J.; Tombler, T. W.; Zhang, Y.; Dai H. *Appl. Phys. Lett.* 2001, 79, 2258-2260.
[106] Robinson, J. A.; Snow, E. S.; Perkins, F. K. *Sens. Actuators*. A 2007, 135, 309-314.
[107] Robinson, J. A.; Snow, E. S.; Badescu, S. C.; Reinecke, T. L.; Perkins, F. K. *Nano Lett.* 2006, 6, 1747-1751.

Reviewed by Haijun Qian, Ph.D., Research Associate, Clemson University Electron Microscope Facility, Clemson Research Park, AMRL Building, 91 Technology Drive, Anderson, SC 29625; Phone: 864-656-6561; Fax: 864-656-2466 ; E-mail: haijunq@clemson.edu.

In: Capacitors
Editor: Kristofer N. Muller
ISBN: 978-1-62417-585-5

Chapter 3

CHARGE TRANSPORT AND BREAKDOWN MECHANISMS OF HIGH-K MATERIALS IN HIGH-DENSITY CAPACITORS

K. B. Jinesh[1,*], W. F. A Besling[1] and J. H. Klootwijk[2]
[1]NXP Semiconductors, Eindhoven, The Netherlands
[2]Philips Research, Eindhoven, The Netherlands

ABSTRACT

The lateral and vertical shrinkage of components in integrated circuits causes abrupt increase in the power consumption of the devices due to the tunneling of charge carriers across the dielectric layer. Novel materials to replace the conventional SiO_2 as the gate electrode have been actively researched in the last two decades, with the advantage of reducing the leakage currents by keeping the capacitive same as that of ultra-thin (<1 nm) SiO_2 films.

This chapter is an overview of the leakage current mechanisms of ultra-thin gate dielectrics used in new-age capacitors and transistors, and their reliability issues.

* Corresponding author, Email: kbjinesh@gmail.com

1. Introduction

The technological roadmaps are based on Gordon Moore's law, which states that the number of devices or components on an integrated circuit (IC) will double every 18 months [1,2]. This aggressive downscaling causes vertical and horizontal shrinkage in device dimensions, and an aftereffect of it is thinning of the gate dielectric layer. When the gate dielectric layer thins down, the electrons undergo quantum mechanical tunneling through the gate, increasing the leakage current abruptly. Leakage current dissipates the power as heat, thus heating up the whole chip, which can lead to the failure of different active and passive components on the chip. The tolerance limit of the leakage through a MOS device is 1 A/cm^2 as per ITRS specifications, which comes in the nano-Ampere range for the structures of the forthcoming CMOS technological nodes. For an ideal dielectric layer (without traps), leakage currents due to tunnelling (direct tunnelling or Fowler-Nordheim tunnelling) is the thickness limiting factor [3,4,5].

1.1. Effect of Dielectric Thinning on Leakage Current

A capacitor is simply a dielectric material sandwiched between two (metallic) electrodes, usually in metal-insulator-metal (MIM) or metal-insulator-semiconductor (MIS) configuration. Ideally, opposite charges accumulate on the electrodes under an applied voltage across it. In an ideal capacitor, the charge density abruptly drops at the electrode-dielectric interface, as illustrated in Figure 1(a), while in a non-ideal capacitor, the leakage mechanisms (tunneling, Schottky emissions, defects etc.) causes a steady current flow (I), which is responsible for the power consumption in the devices ($P=I^2R$).

1.2. Leakage Current Mechanisms in Dielectrics

Let us consider the case when a voltage is applied across the electrodes of a MIS device, the majority of electrons accumulate at the metal-oxide interface and mirror charges (holes) accumulate at the oxide-silicon interface. This image force along the dielectric thickness is given by $q^2/16\pi\varepsilon x^2$, where ε is the dielectric permittivity of the oxide.

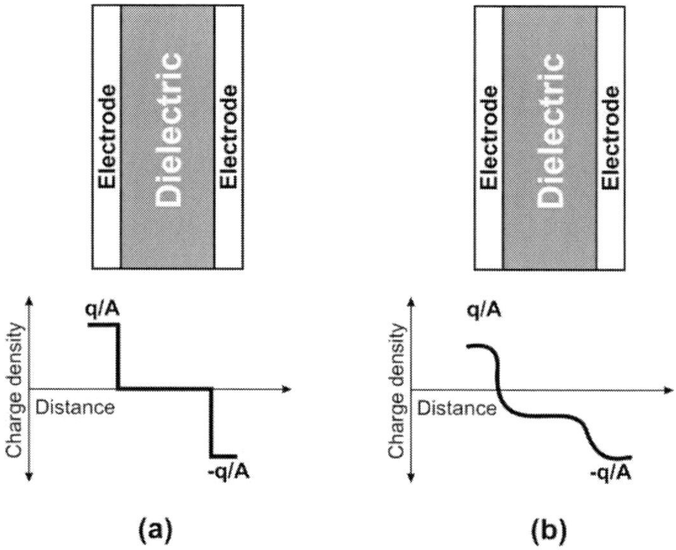

Figure 1. (a) an ideal capacitor with no leakage current across the electrode. The charges are separated by the dielectric medium and the capacitance is simply given by q/V. (b) An instance of a non-ideal capacitor with a non-zero leakage (electron) current. The exact shape of the curve and the extent of steady-state charge density in the dielectric depend on the charge transport mechanism and the material.

At flat-band condition, the free electrons in the conduction band of the metal have to traverse the energy barrier, which is the difference between the conduction band minimum of the dielectric and the Fermi levels of the electrode $\phi_{b0} = E_{C\text{-}Oxide}$- $E_{F\text{-}metal}$ in order to reach the conduction band of the silicon. Due to the image force of the mirror charges, the conduction band of the oxide bends towards the silicon conduction band, as shown in Figure 3.

This phenomenon is generally called Schottky barrier lowering [6,7]. This barrier lowering, $\Delta\phi_b$, is a function of the electric field and is given by $q\Delta\phi_b = q\phi_{b0} - q\phi_b = (q^3E_{ox}/4\pi\varepsilon)^{1/2}$, where E_{ox} is the electric field inside the oxide.

The actual picture of the conduction band is even more complex if one includes defects levels also associated with it. With increasing bias voltage, different charge transport mechanisms emerge. At sufficiently low bias voltages (corresponding to 0 to ~1 MV/cm) when the band bending is negligible, a very minute fraction of the available electrons in the conduction band of the metal resistively passes through the oxide. Also, the very few electrons available in the dielectric medium also contribute to a drift current under the electric field.

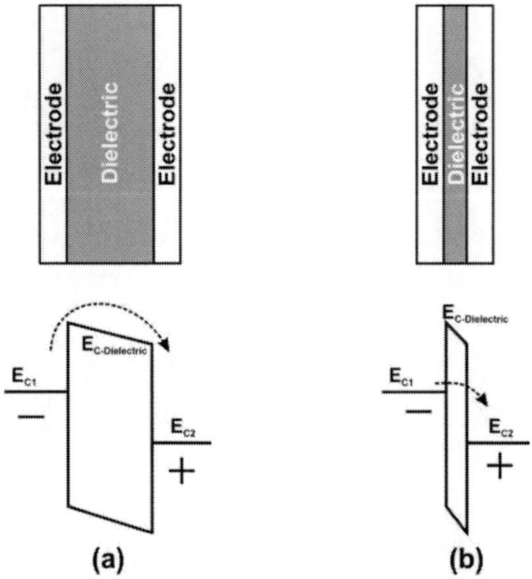

Figure 2. (a) an ideal capacitor, where the leakage current occurs only due to Schottky emission of electrons, when the electrons with sufficient thermal energy cross over the Energy barrier height. (b) In a capacitor with a very thin dielectric, the charges tunnel through the dielectric barrier, causing a drastic increase in the leakage current.

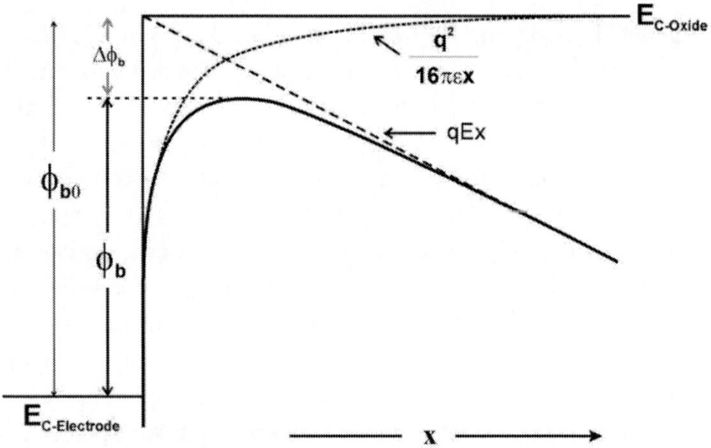

Figure 3. Schottky barrier lowering due to image force under an applied bias across the electrode. The thick line is the resulting conduction band of the oxide under an applied bias, which is the sum of the actual conduction band, the image force and the applied electric field (= qEx).

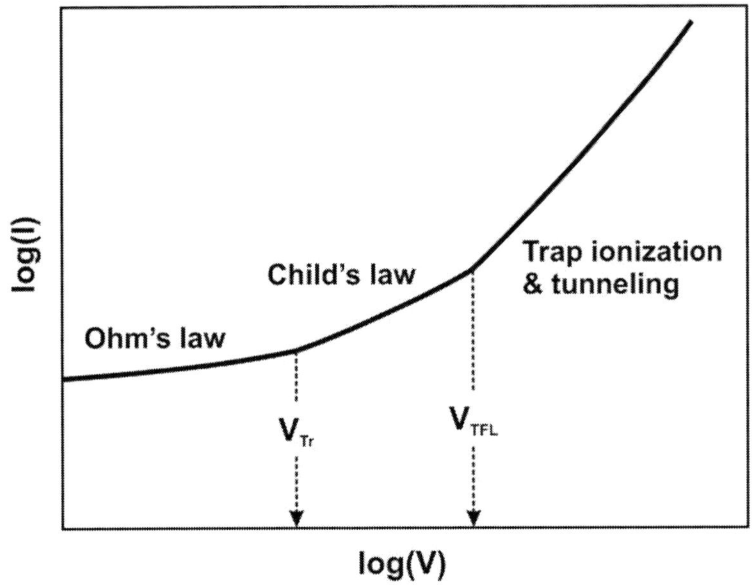

Figure 4. I-V curve in log-log plot showing different conduction regimes that follow different voltage power laws.

This current through the oxide follows Ohm's law and is proportional to the bias voltage ($I_{Ohm} \propto V$).

Increasing the voltage further causes the electrons to trap in the oxide, whose electric field limits the further flow of electrons. The current in this regime is termed as space-charge limited current (SCLC) and follows Child's law [8], which states that the leakage current is proportional to the square of the bias voltage and inversely proportional to the cube of the dielectric thickness ($I_{Child} \propto V^2/d^3$).

The leakage current in the SCLC regime is sensitive to the temperature, because the *shallow* traps liberate free electrons to the conduction band, which is a thermally activated process that follows Arrhenius law. Therefore, SCLC regime is the correct bias range for the temperature-dependent leakage current measurements to draw Arrhenius plots, since the trap levels are filled in these bias voltages. At even higher voltages, the charges trapped in the oxide begin to de-ionize and start contributing to even higher leakage currents through the dielectric.

Simply plotting the voltage-current (*I-V*) characteristics in *log(I) – log(V)* form will reveal the different conduction mechanisms at different voltage regimes. Or, from the slopes of the *log(I) – log(V)* plot, these different regimes of conduction are clearly identified as Figure 4 illustrates. Ohm's law is followed only by the drift electrons in the conduction band of the dielectric (appear due to thermal excitations etc.), which contributes to only a very low leakage, if the dielectric is trap-free. As the voltage increases, more electrons traverse across the dielectric and are trapped in the defects present in the dielectric layer. The electron-filled traps create their own electric field and thus a sphere of influence, which hinder further electrons entering the dielectric. Thus a steady-state current flow is maintained through the dielectric medium, called space-charge limited current (SCLC). The term V_{Tr} in Figure 4 represents the voltage at which the transition from Ohmic (drift) conduction to SCLC regime occurs. From the *trap-filled limit voltage* (V_{TFL}), i.e. the voltage at which the transition from the SCLC regime to trap-free conduction occurs, the trap concentration (N_t) can be estimated as

$$V_{TFL} = qN_t d^2 / 2\varepsilon_r \tag{1}$$

Here, ε_r is the *dynamic* dielectric constant and d is the thickness of the film [9]. The current in the SCLC regime follows the Child's law where the current density is given by

$$J_{Child} = 9\mu\varepsilon_0\varepsilon_r V^2 / 8d^3 \tag{2}$$

Different from Ohm's law, Child's law predicts that the current is proportional to V^2. Thus, the intersect of the SCLC regime in the J-V plot in Figure 4 gives the electronic mobility μ. Using this value, the free carrier density n_0 at room temperature can be calculated from the Ohmic current in the first regime as

$$J_{Ohm} = qn_0\mu V / d \tag{3}$$

When the bias voltage rises beyond V_{TFL}, the charge carriers undergo trap-free conduction and the traps begin to ionize, liberating electrons to the conduction band of the oxide. At electric fields typically lower than ~3 MV/cm, the leakage current is resulting from the electrons that have sufficient energy to cross the metal-oxide barrier ϕ_b. This thermal electron emission resembles the thermionic emission of electrons from a metallic surface at high temperature and therefore, they follow the Schottky thermionic emission given by

$$J_S = AT^2 \exp\left(\frac{-\phi_b + q\sqrt{qE/4\pi\varepsilon_0\varepsilon_r}}{kT}\right) \tag{4}$$

where ϕ_b is the Schottky barrier height and A is the effective Richardson constant [8]. The intersect of the data on the $\ln(J_s/KT^2)$ axis is given by ϕ_b/KT, from which the effective barrier height ϕ_b can be estimated. But, the conduction band of the oxide bends at sufficiently high electric fields (beyond V_{TFL}), as illustrated in Figure 3, and the charge carriers (electrons through the conduction band or holes through the valence band of the oxide) tunnel from the conduction band of the metal to the trap levels in the oxide band-gap and thermally hop through the trap levels to reach the silicon valence band.

This electric-field-assisted thermal hopping of charges is known as Poole-Frenkel (FP) charge injection. The current density due to PF charge injection is therefore similar to the Schottky emission current density, and can be written as

$$J_{PF} = CE \exp\left(\frac{-\phi_t + q\sqrt{qE/4\pi\varepsilon_0\varepsilon_r}}{kT}\right) \tag{5}$$

Here, C is a constant and ϕ_t is the ionization energy of the deep traps. Generally in modelling Poole-Frenkel charge injection, $\Delta E = \phi_t$, where ΔE is the activation energy of the trap, defined by the Arrhenius law described later in this chapter. An example of PF behaviour exhibited by MIS capacitor using $La_yZr_{1-y}O_x$ layer with different La:Zr ratio as the dielectric layer is shown in Figure 5.

Practically, there are number of imperfections and defects in the dielectric materials, due to inhomogenities in deposition process, intrinsic defects in the material during the film growth, dangling bonds at the dielectric/silicon interfaces etc. These create defect energy levels that are distributed over the band-gap of the silicon substrate.

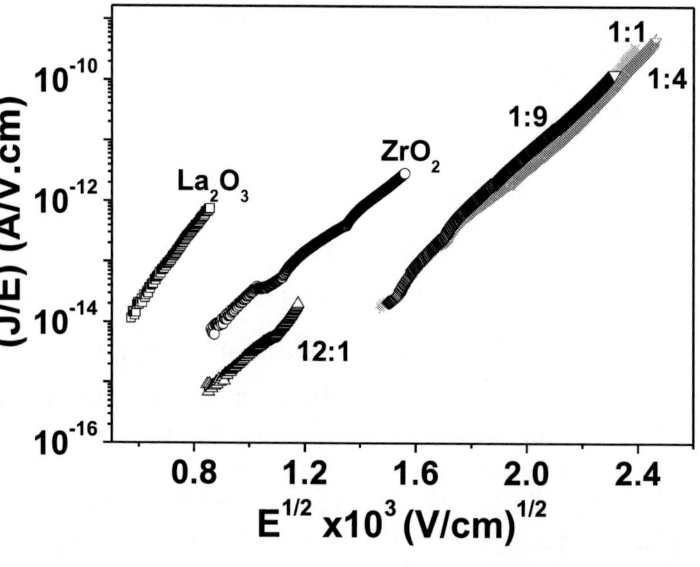

Figure 5. Poole-Frenkel plot of TiN/$La_yZr_{1-y}O_x$/Si devices. For clarity, only the linear section of the J-E curve is shown here. Both the thermal activation energy and dynamic dielectric permittivity of the dielectric layers can be extracted from the slope of these curves.

The lattice mismatching and bonding imperfections at the silicon/oxide interfaces do have effects on the band diagram of the oxide as well. Trap levels due to the intrinsic defects in the oxide layer, such as vacancies and grain boundaries, appear as shallow trap levels close to the conduction band of the oxide and they are thermally excitable electronic states. Since these trap levels are close to the conduction band of the oxide, there is a non-zero probability that the wave-function of these electrons overlap with the conduction band and appear as free electrons.

The probability of these electrons to traverse the energy barrier to reach the conduction band exponentially increases with the temperature of the system.

This thermally activated process can be described using rate equation, where the rate of successful jumps of electrons from the trap levels to the conduction band depends on the energy barrier height ΔE and the temperature of the ensemble by the expression [10]

$$r = r_0 \exp(-\frac{\Delta E}{k_B T}) \tag{6}$$

where r_0 is the attempt frequency of an electron in the potential well shown in Figure 5 (c), r is the rate of successful jumps from the well to the conduction band of the dielectric and k_B is the Boltzmann constant. This expression tells that the number of thermally activated electrons reaching the conduction band and contributing to the leakage current increases with temperature for a given trap level.

Therefore, it is safe to assume that $r_0 \propto J_0$ and $r \propto J$, where J_0 and J are current densities in a-thermal (infinite temperature) and thermally activated regimes of charge conduction. Thus expression (6) becomes

$$J = J_0 \exp(-\frac{\Delta E}{k_B T}) \tag{7}$$

Therefore, the slope of the Arrhenius plot where $ln(J/J_0)$ is plotted against $1/T$ gives the activation energy ΔE of the trap levels. From Eqn. (8), ΔE can be derived independent of J_0 as

$$\Delta E = -\frac{k_B}{e} \frac{\log \frac{J_2}{J_1}}{\left(\frac{1}{T_2} - \frac{1}{T_1} \right)} \tag{8}$$

in units of electron volts (eV). Arrhenius behaviour can be traced by measuring the leakage current at a sufficiently low voltage (such that the estimated activation energy will not be affected by the bending of the conduction band) at different temperatures. As an example, the Arrhenius behaviour of leakage current through 22 nm thick ZrO_2 on Si substrate and TiN top electrode is shown in Figure 7.

Generally, when using the slope of the $ln(J_0)$ versus $1/T$ linear distribution, J_0 is an unimportant parameter. But this parameter gives indications about the Arrhenius behaviour of traps levels. Since the Arrhenius equation (Eqn. 7) deals only with the energy barrier between the trap energy E_t and the conduction band E_C, so that $\Delta E = E_C - E_t$, the pre-exponential factor J_0 is purely the current due to the trapped electrons at infinite temperature, where all the trap levels are emptied by thermally activated electrons.

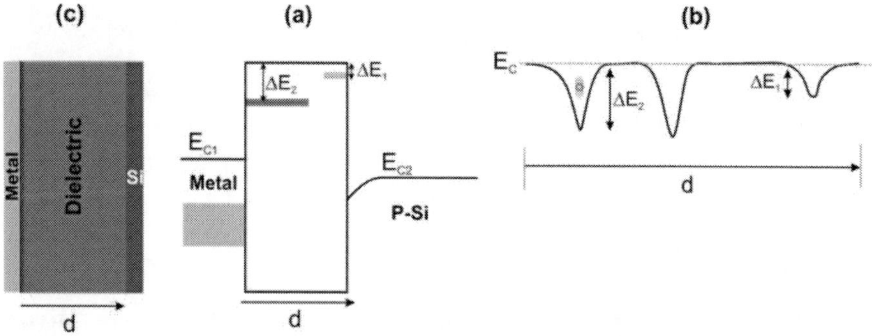

Figure 6. (a) Model of a MIS capacitor on Silicon; (b) Band-diagram of the device with trap levels ΔE_1 and ΔE_2 representing the shallow traps; (c) Trap levels close to the conduction band of the oxide at zero applied bias voltage. Thermal excitation of the electrons is depicted in the first potential well.

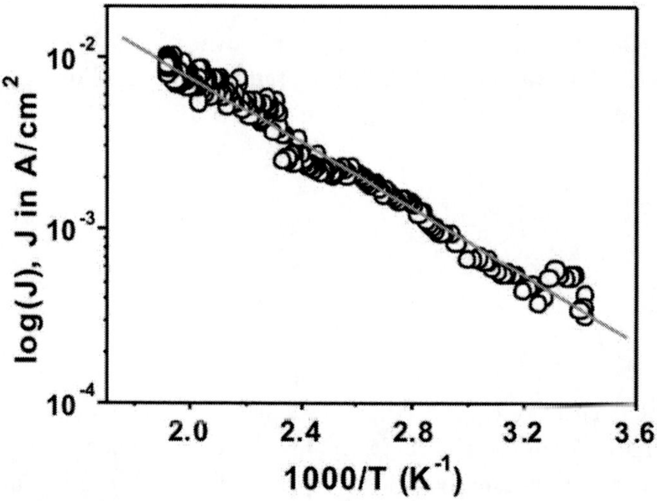

Figure 7. Arrhenius plot of the leakage current through 22 nm ZrO_2 showing a straight line. From the slope of the data, thermal activation energy of 0.19 eV is obtained.

Therefore, J_0 represents the maximum capacity of the trap levels continuously trapping and releasing electrons. Thus J_0 should have similar behaviour as of the fixed traps in the oxide.

It is worth noting that the equations (1), (2), (4) and (5) employ the dynamic dielectric constant ε_r instead of static low-frequency dielectric constant. This is because, in the Schottky emission picture, the electrons reside at the metallic surface only for a very small time. Similarly, the pre-exponential factor in Arrhenius law, which is the attempt frequency of the electrons in a potential well set by the traps, is ~ 10^{13} s^{-1}. This high-frequency dielectric constant (the optical dielectric permittivity) is related to the refractive index of the material by $\eta = \sqrt{\varepsilon_r}$ and thus, the refractive index of the oxide also can be deduced *if* it exhibits PF charge injection [11].

At higher electric fields, the conduction band of the oxide bends towards the anode to the extent that it triggers electron tunnelling through the oxide. This type of charge conduction is called Fowler-Nordheim (FN) tunnelling, and is extensively used in field emission from sharp metallic tips (where the barrier is set up by the resistance of air or vacuum). FN tunnelling is purely a quantum mechanical tunnelling of electrons through a rounded triangular potential energy barrier of the oxide. This tunnelling takes place from the conduction band (or Fermi level if the electrode is a metal) of the cathode through the oxide to the conduction band of the p-type silicon substrate (anode) in the case of electrons and from the valence band of the silicon through the oxide to the valence band of the cathode material for holes. Whether the charge carrier is electrons or holes depends primarily on the conduction and valence band offsets of silicon with that of the oxide. The current density due to FN tunnelling is given by [12]

$$J_{FN} = \frac{q^3 m}{16\pi^2 \hbar m_{ox} \phi_b} E^2 \exp\left(-\frac{4}{3} \frac{\sqrt{2m_{ox}}}{q\hbar E} \phi_b^{3/2} \right) \tag{9}$$

where ϕ_b is the tunnelling barrier height and m_{ox} is the electron effective mass in the oxide. The tunnel barrier height can be estimated from the slope of the FN plots, which is a plot with $ln(J/E^2)$ plotted against $1/E$ yielding a straight line. Fowler-Nordheim tunnelling though the Al_2O_3 dielectric layers deposited by means of thermal and plasma-assisted atomic layer deposition (ALD, Ref. 13) is shown in Figure 8 as an example.

The major conduction mechanisms described in this section are summarised in Figure 9. The main parameter estimated from FN plots is the barrier height ϕ_b (not the trap activation energy ϕ_t), which is the conduction/valence band offset of the dielectric with either of the electrodes. But to estimate the barrier heights accurately, one has to employ an effective electron mass (m^*) in the oxide.

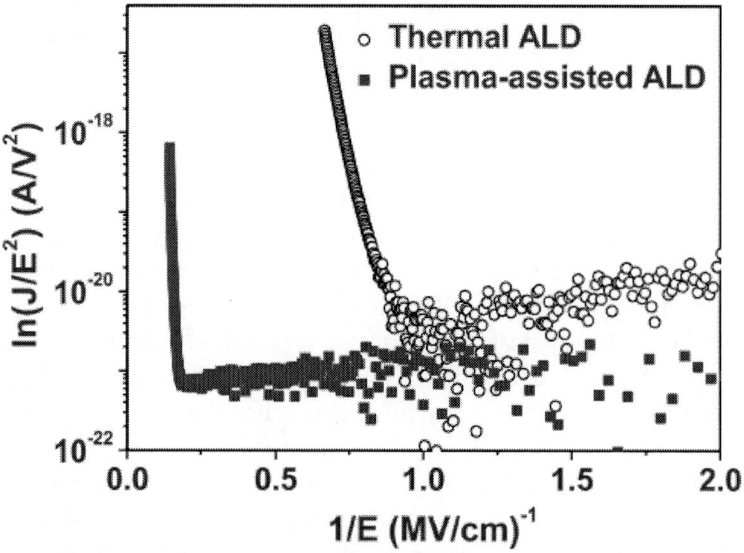

Figure 8. Fowler-Nordheim behaviour of Al/Al_2O_3/Si devices deposited by thermal atomic layer deposition (T-ALD) and plasma-assisted atomic layer deposition (PA-ALD) techniques. The steeper slope of PA-ALD Al_2O_3 shows a larger Al_2O_3-Si barrier height, due to the superior interface formation by the plasma-enhanced process. FN plots explicitly demonstrate the interface quality of the material.

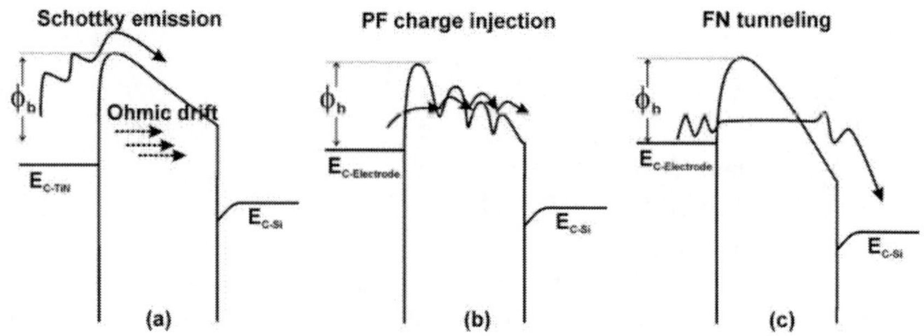

Figure 9. Major conduction mechanisms through the dielectric at (a) low electric fields, exhibiting thermal jumps over the barrier, described by Schottky emission; (b) medium electric field (2-4 MV/cm) showing thermal hopping of charges through defect levels, assisted by electric-field induced bending of the conduction band and (c) high electric fields (>4 MV/cm), where the conduction band is sufficiently deformed to allow easier charge tunnelling through the dielectric medium.

The discussion about the accuracy of barrier heights obtained from the slopes of normal FN plots is rather conflicting because of two reasons. Firstly, the effective mass can be estimated accurately only for a crystalline material, since the apparent mass of an electron arises due to the interaction with a periodic crystal lattice and therefore influenced by the phonon drag and the interaction with other electrons in the system. Most of the dielectric materials are amorphous in nature, or they vary their crystal orientation according to the deposition techniques and growth conditions. Therefore, the value of the effective mass varies in a wide range. For instance, $m^* = 0.15\ m_0$ [11], and 0.26 m_0 [14,15] are reported for La_2O_3 thin films. Second reason is that for a given electric field, the effective barrier height is always lower by a factor of $\sqrt{q^3 E_{ox} / 4\pi\varepsilon_0\varepsilon_r}$ as shown in Figure 3, due to Schottky barrier lowering, which implies that the barrier height decreases continuously with the applied field. There is an additional barrier lowering due to temperature, which is in fact not included in the original FN equation. Figure 10 shows the current density – electric field (J-E) behaviour of a $TiN/Er_2O_3/Si$ device and the fitting using PF, FN conduction mechanisms and their combination.

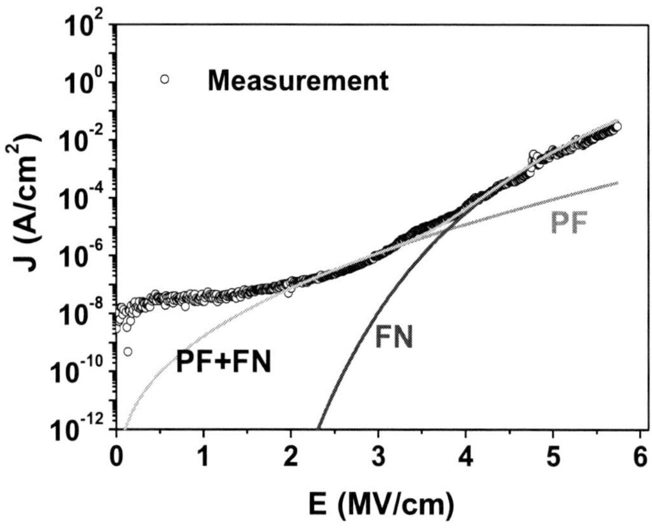

Figure 10. J-E data measured in $TiN/Er_2O_3/Si$ device and theoretical fits with PF and FN tunnelling. The initial part of the J-E data seems to have a much larger current than predicted by PF emission, which is mainly because of the detection limit of the measuring instrument. Also, Ohmic and SCLC contributions to the current are not included in this fitting.

1.3. Dielectric Breakdown in Capacitors

The breakdown voltage (V_{BD}) or the breakdown field (E_{BD}) defines the strength of a dielectric and therefore, it directly reflects the reliability of the device. Most of the dielectric materials follow an empirical law between the dielectric breakdown field and the dielectric permittivity (k) given by

$$E_{BD} = \frac{20}{\sqrt{k}} MV / cm \tag{10}$$

Though this is an empirical model, several binary oxides have been shown to obey a similar trend [16], which means that the usage of a dielectric with a higher k-value has the disadvantage of early breakdown. The breakdown behaviours of TiN/$La_yZr_{1-y}O_x$ /Si capacitors with different La:Zr ratio are shown in Figure 11.

Several models have been proposed in the past to understand the breakdown phenomena of oxides under an applied electric field. The hard breakdown of oxides is usually avalanche breakdown induced by the hot electrons or holes passing through the oxide. When electrons pass through the oxide layer under an applied electric field, defects such as traps, interface states, and positively charged donor-like states etc. accumulate gradually in the oxide, until a critical defect density is reached, when the oxide destructively breaks down. This critical defect density is approximately 5×10^{13} /cm^2 in oxides thicker than 5 nm [17]. At these defect densities, nanometre-long conductive filaments – percolation paths – form in the oxide initiating the final breakdown of the device. When the oxide thickness is diminished to the typical defect size for that material, generation of a single defect in the oxide is sufficient to cause electrical breakdown of the entire film [18]. Although earlier computer simulations of Stathis have shown that the typical defect diameter is ~3 nm, later works by Nicollian et al. indicates that the defect size is much smaller, of the order of 0.4 nm [19].

The dominant charge transport mechanism at electric fields larger than ~3 MV/cm is Fowler-Nordheim tunnelling. Theoretically, quantum mechanical tunnelling is a non-destructive mechanism, but it often generates defects in the oxide.

The charge transport from one electrode to the other can be either tunnelling, or transport through the conduction band of the oxide. Tunnelling is an elastic process by definition.

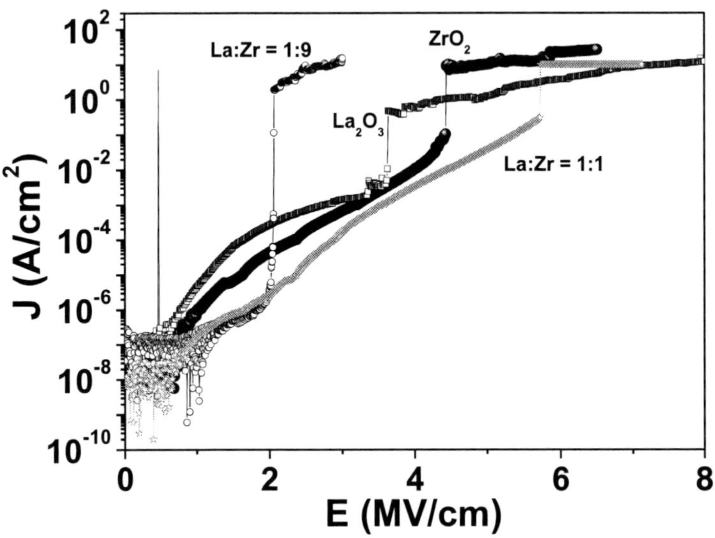

Figure 11. Breakdown behaviour of TiN/$La_yZr_{1-y}O_x$/Si devices showing the influence of Zr content in device failure.

Charge transport through the conduction band can be either elastic or inelastic, or ballistic tunnelling, without losing the energy of the charges. In a ballistic process, the net energy of the charges can be higher than the initial energy when they reach the anode, since the transported carriers do not have collisions with atoms or lattices under an applied electric field.

In an inelastic process, this charge scattering primarily excites the longitudinal phonon modes of the (crystalline) oxide lattice [20,21]. At higher electric fields, the energy gained by the charge carriers by accelerating through the electric field is balanced by the energy lost into the acoustic phonon modes of the oxide.

Apparently, inelastic tunnelling, also known as *steady-state FN tunnelling*, is a dissipative charge transport process that can initiate oxide degradation. However, several breakdown models have been proposed to explain the oxide degradation and time dependent dielectric breakdown (TDDB) experiments are regarded as standard ways to test these models. Two important field-based models in this category are the *1/E* model and the E-model. *1/E* model assumes that the major cause of oxide breakdown is generation of holes and the breakdown occurs when a critical hole-fluence has passed through the oxide [22]. These holes are generated by the impact ionization [20] in the oxide due to tunnelling of electrons through it, or due to the energy release of the tunnelling electrons at the oxide-anode interface. In the latter case, the holes generated at the interface tunnel back to the oxide, degrading the oxide.

The E-model is basically a thermochemical model and was originally an empirical model proposed by Anolick in 1979 [23]. This was later on described on the basis of thermodynamics [24,25]. In this model, the time taken for 50% of the device breakdown (termed as t_{50}) can be expressed as

$$t_{50} = A\exp(-\gamma E)\exp(-\frac{\Delta E}{kT}) \tag{11}$$

Here, γ is the acceleration parameter in the accelerated breakdown experiments (the slope of the log(t_{BD})-E curves), A is a constant, E is the electric field in the oxide, k is the Boltzmann constant and ΔE is the activation energy of defect generation.

The electric field models are important to estimate the reliability of the devices in the accelerated tests usually termed as time-dependent dielectric breakdown (TDDB) tests. As mentioned before, the *1/E*-models of oxide breakdown assume that dielectric breakdown is a current driven process, caused by tunnelling of the charge carriers such as hot holes injected from the anode (called the Anode Hole Injection model – AHI) due to the energy release of tunnelling electrons at the oxide-anode interface [26]. Since the current driving mechanism is mainly FN tunnelling, the expression for t_{50} has a *1/E* dependence and is given by

$$t_{50} = \tau_0 \exp(\frac{G}{E})\exp(-\frac{\Delta E}{kT}) \tag{12}$$

where τ_0 and G are constants. [27,28].

1.4. Time Dependent Dielectric Breakdown in Dielectric Thin Films

As discussed in the previous section, the dielectric breakdown occurs when the defect density in an oxide reaches a critical value. When the electric field in the oxide is high enough, defects accumulate over time and the device finally breaks down. At a given voltage across the electrodes, the time until the oxide breaks down is known as time-to-breakdown (t_{BD}) and the critical defect density for the final breakdown of the device is explicitly linked to the total charge passed through the oxide layer until it breaks down destructively. This total charge passed through the device until the critical defect density is reached is known as the charge-to-breakdown or Q_{BD} given by

$$Q_{BD} = \int_0^{t_{BD}} I(t)dt \tag{13}$$

where $I(t)$ is the leakage current measured at the fixed stress voltage. Q_{BD} is usually determined by applying a fixed voltage across the capacitor electrodes and measuring the stress current through the device until it breaks down. This method is generally known as constant voltage stress (CVS) measurement.

Generally, Q_{BD} has been shown to have strong thickness dependence below 5 nm for silicon oxide, where the electric field models of breakdown do not apply [29], and has strong polarity dependence suggesting that the breakdown in the oxides is determined by the applied voltage-driven maximum energy of the charge carriers, not their electric-field-driven average energy [30]. For the same reason, Q_{BD} has a strong dependence on the applied voltage across the electrodes. At larger magnitude of the leakage currents (at voltages close to the breakdown voltage), the number of electrons with sufficient energy to induce defects in the oxide are large enough to initiate the avalanche breakdown in the oxide and measurement at these voltages give erroneous Q_{BD} values. The correct estimation of this parameter would be at voltage low enough to induce defect accumulation instead of avalanche breakdowns.

Therefore, if Q_{BD} is plotted against the leakage current density (J), the data points are distributed in two branches, one where the avalanche breakdown dominates and Q_{BD} vary with J, and another branch where Q_{BD} is steady and independent of J. Figure 12 shows the leakage current through $La_yZr_{1-y}O_x$ films with different La:Zr ratio in the film, measured at the same electric field of 5MV/cm.

If CVS measurements are carried out at different voltage across the devices and plot the Q_{BD} values against the average leakage current density at the corresponding voltages, a plateau of constant Q_{BD} values obtained at voltages below certain threshold, as Figure 13 shows. It suggests that the CVS measurement should be carried out at a range of sufficiently low voltages, to obtain the correct Q_{BD} values of that dielectric material.

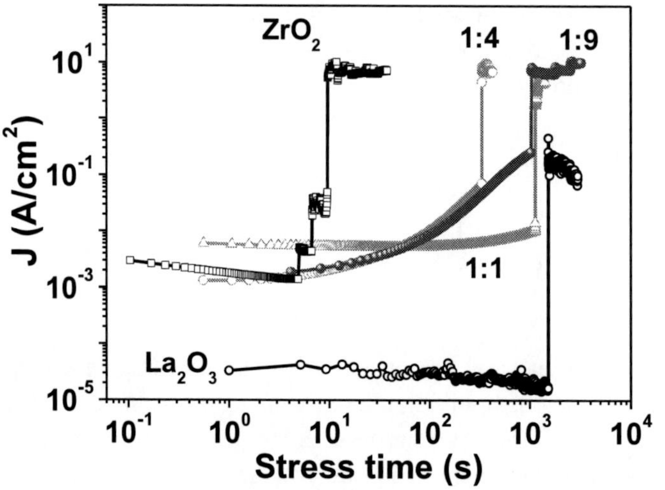

Figure 12. Stress current measured for different $La_yZr_{1-y}O_x$ films at the same electric field of 5 MV/cm. Q_{BD} is estimated by integrating the area under the curve until (the first) breakdown point.

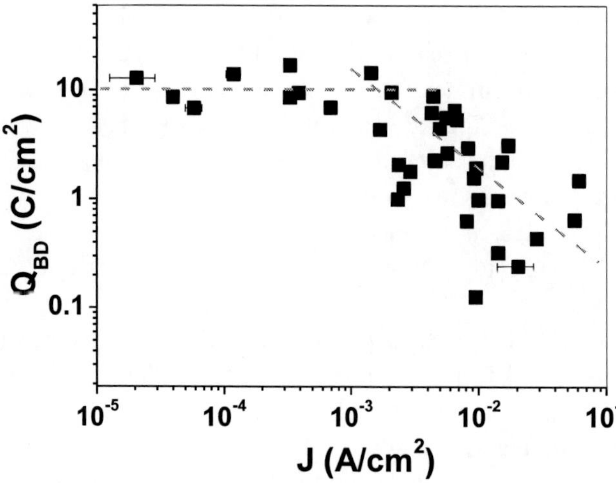

Figure 13. Distribution of Q_{BD} against the leakage current (J) at different voltages measured for ALD processed Al_2O_3.

Trap generation is a statistical process, and it is known that the generation of neutral electron traps during high field stress is directly linked to the neutral electron trap generation during high field stress [31].

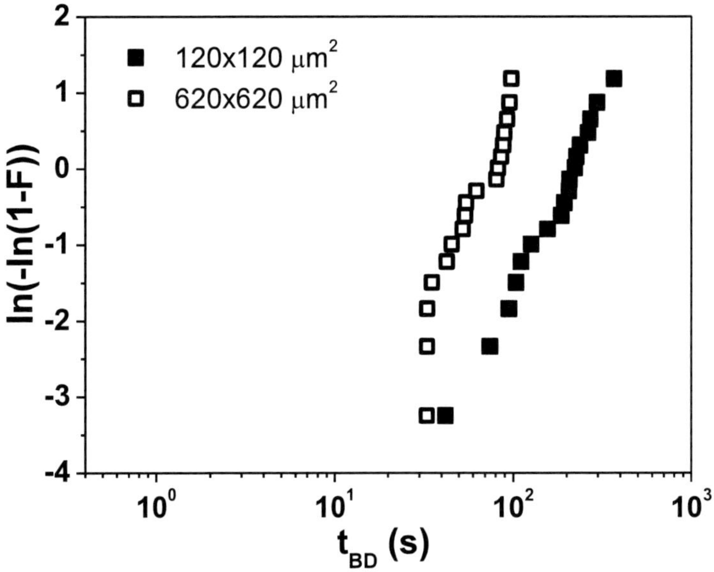

Figure 14. Weibull distribution of t_{BD} values for two different TiN electrode areas, measured for TiN injection condition at 8V. Further measurements reported in this Chapter were carried out on devices with area 120×120 μm^2.

This section investigates the dielectric strength of the $SiO_2/La_yZr_{1-y}O_x$ stacks using lifetime extrapolation measurements. The important parameter is time-to-breakdown, denoted at t_{BD} and is simply the time taken since the voltage is applied until the breakdown of the dielectric. Since the defect formation in the dielectric stack is related to the applied bias voltage across it (due to the FN tunnelling of the electrons through the conduction band of the high-*k* layer), the time-to-breakdown (t_{BD}) is measured in this technique as a function of the bias voltage. Due to the statistical nature of defect generation under the applied bias across the stack, the measured t_{BD} values depend on the top electrode area of the capacitor. This is illustrated in Figure 14 by measuring the t_{BD} values of 13.3 nm $La_yZr_{1-y}O_x/SiO_2$ stack for two different electrode areas.The larger the voltage applied across the dielectric layer, the faster is the breakdown process. Thus the exponential breakdown model suggests that t_{BD} varies exponentially with the applied voltage, which follows the relationship [32]:

$$t_{BD} = t_0 \exp(-\gamma V) \tag{14}$$

where γ is known as the voltage acceleration factor and V is the applied voltage across the oxide. As per this expression, t_{BD} has an exponential relation with the voltage. Figure 15 shows the dependence of t_{BD} with bias voltage, for electron injection from Si and TiN.

Using Eqn. (14), the acceleration parameter γ can be calculated for different thickness as plotted in Figure 16.

For very thin films (~1.5 nm) of HfSiON/poly-Si stack, acceleration factor of γ ~ 7 decade/V has been reported (Ref. [33], page 53). However, it should be noted that the exponential model given by Eqn. (14) is applicable only for thicker films so that the interface defect density will not influence the degradation rate considerably. This explains the linear decrease of the acceleration factor with the film thickness. On the other hand, acceleration parameter γ is a material parameter that is independent of the film thickness, and can be written as [34]

$$\gamma = -\left(\frac{\partial \ln t_{BD}}{\partial E}\right) = \frac{\Delta P}{k_B T} \tag{15}$$

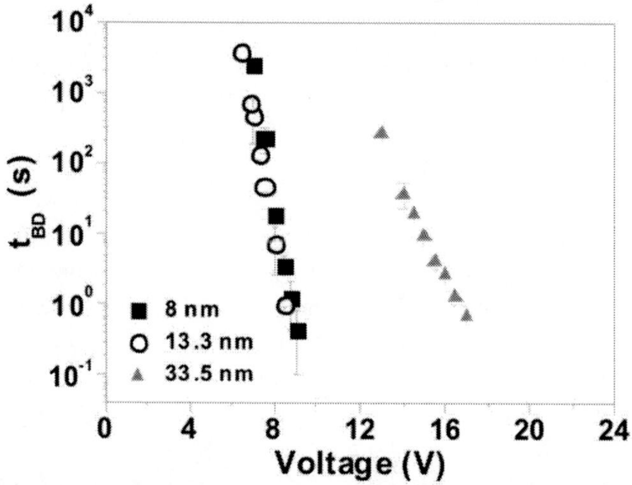

Figure 15. t_{BD} values plotted as a function of voltage for the devices with different thicknesses of $La_yZr_{1-y}O_x$ layer. Capacitors with area of 120×120 μm^2 were used for these measurements.

where, E is the electric field and ΔP is the change in polarization of the dielectric system from a stable to breakdown transition state.

The power law model gives a slightly more optimistic extrapolation compared to the exponential model given by Eq. 14. Figure 17 shows the power-law dependence of t_{BD} with voltage measured for TiN/$La_yZr_{1-y}O_x$/Si devices.

In fact, by knowing the n value of a given dielectric layer with a given thickness, we can extrapolate the voltage for a desired "guaranteed" time for safe operation of the device. This is generally indicated as 10-year operational voltage of the device and is a technologically relevant parameter of the capacitor.

Another method for the lifetime extrapolation of dielectric films for different voltages is the *voltage power law model* given by [35]

$$t_{BD} \propto V^{-n} \tag{16}$$

Figure 18 gives the 10-year operational voltage of the TiN/$La_yZr_{1-y}O_x$/Si devices, measured when TiN and Si are biased (and the other electrode is grounded).

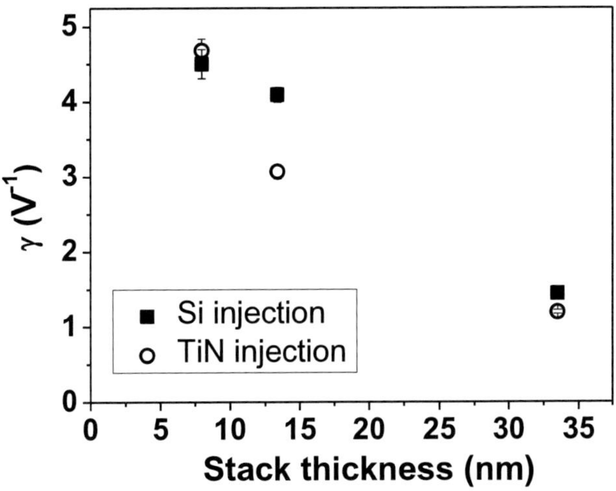

Figure 16. The acceleration factor γ measured for TiN/$La_yZr_{1-y}O_x$/Si devices, as a function of the $La_yZr_{1-y}O_x$ thickness, measured from both TiN and Si electrodes.

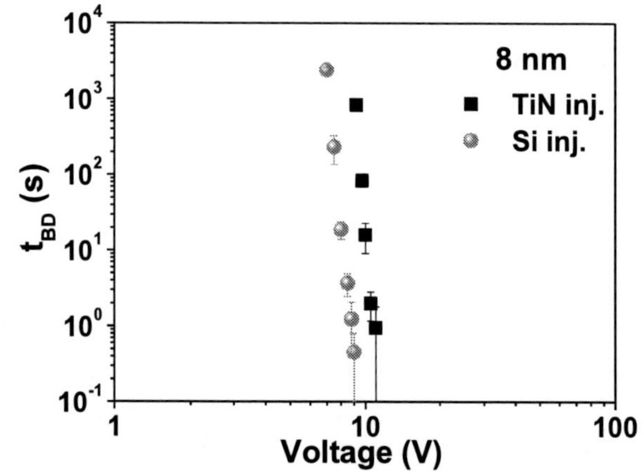

Figure 17. T_{BD} plotted against voltage for the TiN/$La_yZr_{1-y}O_x$/Si capacitors, measured also when the electric field direction is inverted. The slope of the data is the power n in Eqn. (16).

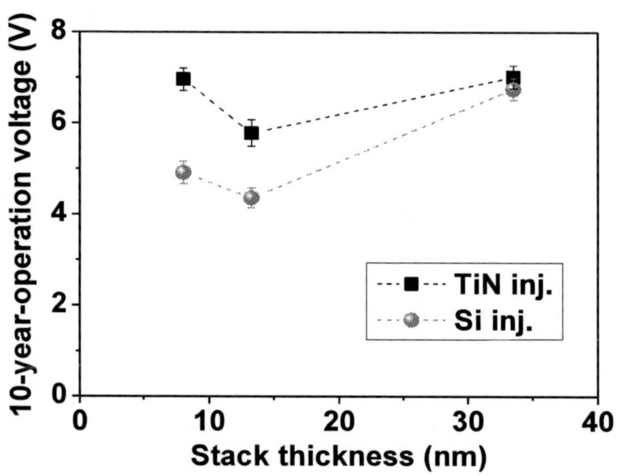

Figure 18. Extrapolation of 10-year operational voltage of the TiN/$La_yZr_{1-y}O_x$/Si stacks as a function of the stack thickness.

Conclusion

This chapter describes about the general mechanisms that are relevant to describe the leakage current through capacitors with thin dielectrics films. The thermally assisted Poole-Frenkel charge injection and field-induced tunneling,

Fowler-Nordheim tunneling, are the major conduction mechanisms in capacitors. Since both the injected electrons and the hot holes created by them can degrade the dielectric films, the general mechanisms related to the dielectric reliability also are described in this chapter, thus introducing the charge-to-breakdown parameter as a standard measure for the dielectric strength of a material.

REFERENCES

[1] G. Moore, *Electronics*, 38, 114 (1965).
[2] G.E. Moore, *Tech. Dig. IEEE Int. Electron Device Meeting*, page11, (1975).
[3] M. Lenzlinger, E.H. Snow, *J. Appl. Phys.*, 40, 278 (1969).
[4] G. Pananakakis, et al., *J. Appl. Phys.*, 78, 2635 (1995).
[5] H. C. Lin, P. D. Ye, G. D. Wilk, *Solid-State Electron.*, 50, 1012 (2006).
[6] S.M. Sze, Physics of Semiconductor Devices, Ed.2, Wiley Publications, New York, 198.
[7] J.H. Klootwijk, Interpoly dielectrics for Non-volatile Memories, PhD Thesis, University of Twente, 1997.
[8] J.G. Simmons, *J. Phys. D. Appl. Phys.*, 4, 613 (1971).
[9] F.C. Chiu, H.W. Chou, J.Y. Lee, *J. Appl. Phys.*, 97, 103503 (2005).
[10] S.L. Miller et al., *J. Appl. Phys.*, 70, 4555 (1991).
[11] S. Ramanathan, C.-M. Park, P.C. McIntyre, *J. Appl. Phys.*, 91, 4521 (2002).
[12] R.H. Fowler, L. Nordheim, Proceedings of the Royal Society of London, 119, 173 (1928).
[13] K.B. Jinesh et.al.,*J. Electrochem. Soc.,* 158, G21 (2011).
[14] E. Miranda, J. Molina, Y. Kim, H. Iwai, *J. Non-Crystalline Solids,* 352, 92 (2006).
[15] Y. Yeo, T. King, C. Hu, *Appl. Phys. Lett.*, 81, 2091 (2002).
[16] P. Jain, E. J. Rymaszewski, *IEEE Transactions on Advanced Packaging*, 25, 454 (2002).
[17] D.J. DiMaria, E. Cartier, D. Arnold, *J. Appl. Phys.*, 73, 3367 (1993).
[18] J.H. Stathis, *J. Appl. Phys.*, 86, 5757 (1999).
[19] P.E. Nicollian et al., *IEDM Technical Digest*, 743 (2006).
[20] T.H. DiStefano, M. Shatzkes, *Appl. Phys. Lett.*, 25, 685 (1974).
[21] W.T. Lynch, *J. Appl. Phys.*, 46, 998 (1975).

[22] I.C. Chen *et al, Appl. Phys. Lett.*, 49, 669 (1996).
[23] E. S. Anolick and G. Nelson, Proceedings of the IEEE-IRPS (IEEE, NewYork, 1979), Vol. 7, page. 8.
[24] J. W. McPherson, D. A. Baglee, Proceedings of the IEEE-IRPS (IEEE, New York, 1985), Vol. 23, p. 1.
[25] J. W. McPherson, D. A. Baglee, *J. Electrochem. Soc.*, 132, 1903 (1985).
[26] K.F. Schuegraph, C. Hu, *Proceedings International Reliability Physics Symposium (IRPS)*, P.126, 1994.
[27] J. W. McPherson, H. C. Mogul, *J. Appl. Phys.*, 84, 1513 (1998).
[28] S-C. Chang, S.-Y. Deng, and J. Ya-M. Lee. *Appl. Phys. Lett.*, 89, 053504 (2006).
[29] G.M. Paulzen, *Microelectronic Engineering,* 36, 321 (1997).
[30] P.E. Nicollian, Physics of Trap generation and Electrical breakdown in Ultra-thin SiO_2 and SiON gate dielectric materials, PhD Thesis, Univerity of Twente, 2007
[31] R. Degraeve et al., *IEEE Trans. Electron Devices*, 45, 904 (1998).
[32] T. Nigam, *Growth* kinetics, electrical characterization and reliability study of sub-5nm gate dielectrics, PhD Thesis, Catholic University, Leuven (1999).
[33] T. Kauerauf, Degradation and Breakdown of MOS gate stacks with High permittivity dielectrics, PhD Thesis, Catholic University, Leuven (2007).
[34] H. Cui, P.A. Burke, *Appl. Phys. Lett.*, 84, 2629 (2004).
[35] E.Y. Wu, et al., *IEDM Technical Digest*, p. 541-544 (2000).

In: Capacitors
Editor: Kristofer N. Muller

ISBN: 978-1-62417-585-5

Chapter 4

INTEGRATION OF CAPACITORS INTO INTEGRATED CIRCUITS: BEYOND MOORE'S LAW

K. B. Jinesh[1,*], J. H. Klootwijk[2] and W. F. A. Besling[1]
[1]NXP Semiconductors, Eindhoven, The Netherlands
[2]Philips Research High Tech, Eindhoven, The Netherlands

ABSTRACT

While Integrated circuits consume only 10% of the volume of a circuit-board, the majority of the space is occupied by passive elements such as capacitors, inductors or resistors. At system-level, integration of passive elements plays a crucial role in the miniaturization of functional devices such as mobile phones, computers and i-pads. However, integration of passive elements is technologically challenging, because some of these elements, for instance capacitors, can not compromise the reduction in the area of the device. Several smart technological innovations have been devised in the last decade to circumvent the "available reducing area in a chip", and this chapter gives an overview of the technology of integrated capacitors and their functioning.

[*] Corresponding author, E-mail: kbjinesh@gmail.com.

1.1. INTRODUCTION

Gordon Moore predicted in 1965 that the active components on an integrated circuit (IC) will double every 18 months to cope up with the future technological needs [1, 2]. Cramming more components into an IC results in the lateral and vertical shrinkage of the dimension of individual components in it, which poses several challenges including large leakage currents and low reliability of the devices. The International Technology Roadmap for Semiconductors (ITRS) foresees several challenges in complying with Moore's law in future semiconductor devices [3], among which two reasons are worth noting. Firstly, Hirose et al. calculated the tunnel current through the dielectric layers based on multiple-scattering theory, which suggests that the transconductance fluctuations will be problematic when the gate oxide thickness is scaled down to 0.8 nm [4]. This sets the fundamental limit for thinning down the gate oxide (SiO_2) in complementary metal-oxide-semiconductor (CMOS) devices to 0.7 nm, beyond which the spill-over of the silicon conduction-band wave-functions into the oxide generates interface states that will generate unacceptably large power consumption [5]. Secondly, every bit of data processed in a CMOS logic gate consists of a thermodynamically irreversible process, which is given by the Landauer limit $k_B T \ln 2$ = 17.9 meV [6], which might set the noise limit beyond which the CMOS technology may not be possible unless another smart technology is developed [7].

In fact, the technological challenges in the modern world is not set by Moore's law alone; though Moore's original paper mentions about the importance of passive integration as well, the commonly accepted Moore's law deals mainly with the downscaling of the active elements inside integrated circuits (ICs). The microchips that follow Moore's law are only 10 percent of the systems, where the rest consists of large passive components such as capacitors, inductors, sensors, antennas, resistors etc. Therefore, the passive integration has a tremendous technological impact on component-level integration [8]. Technologists are currently looking towards the system-level integration of devices such as cell phones, digital television receivers, health monitoring systems and sensors, which are functioning based on a different technological grounds. Integration of large systems into one system has led to two complementary directions. One is the "*More Moore*" concept with the urge towards extremely small device dimensions and thus the development of system-on-chip (SoC) concepts. Second is the "*More than Moore*" concept, pursuing the integration of the passive elements that are excluded from the domain of Moore's law, which lead to the System-in-Package (SiP) solution [9]. Philips has launched a series of highly miniaturized RF transceiver modules with Si-based SiP technology [10] employing back-end processing and flip-chip hetero-integration with RF (Bi)-CMOS technology and assembly in a package [11].

Integration of passive elements, especially capacitors is not technologically straightforward. Let us consider the standard expression for the capacitance of a parallel plate capacitor given by [12]:

$$C = \frac{\varepsilon_0 k A}{d} \qquad (1)$$

Here, ε_0 is the dielectric permittivity of free space, k is the dielectric constant of the dielectric medium and A is the area of the capacitor and d is the thickness of the dielectric medium. From this simple equation, it is clear that the capacitance can be increased by 1) increasing the dielectric constant k, 2) increasing the area of electrodes A, and 3) reducing the thickness d of the dielectric layer. Here the challenges of integrating the capacitors into the *More than Moore* concept are the following: firstly, increasing the area within a chip is normally nonnegotiable, since the total silicon area on a chip is reducing and the number of component processed over it is increasing every year as per Moore's law.

Secondly, reducing the dielectric thickness has certain limits. The probability of the charge carriers (either electrons or holes) to quantum mechanically tunnelling through the dielectric medium exponentially increases with reducing the dielectric thickness. This causes increase in leakage currents as the thickness reduces. Leakage current dissipates power as heat, thus heating up the whole chip, which can lead to the failure of different active and passive components on the chip. The tolerance limit of the leakage through a MOS device is 1 A/cm^2 as per ITRS specifications. For the thickness limit of an ideal dielectric layer (without traps), leakage currents due to tunnelling (direct or Fowler-Nordheim) is the determining factor [13, 14].

1.2. Tackling the Area Problem: Substrate Patterning

In order to enhance the area available in a chip, the idea of micro-patterned silicon has been put forward by *Lehmann* et al [15], where deep pores with large aspect ratios (10-20) are etched in silicon either by deep reactive ion etching (DRIE) or by wet chemical etching. By forming an array of such deep pores etched in silicon, the total area of the capacitor can be enhanced by 10-15 times [16]. Figure 1 shows an example of such micro-pores etched in silicon using DRIE process.

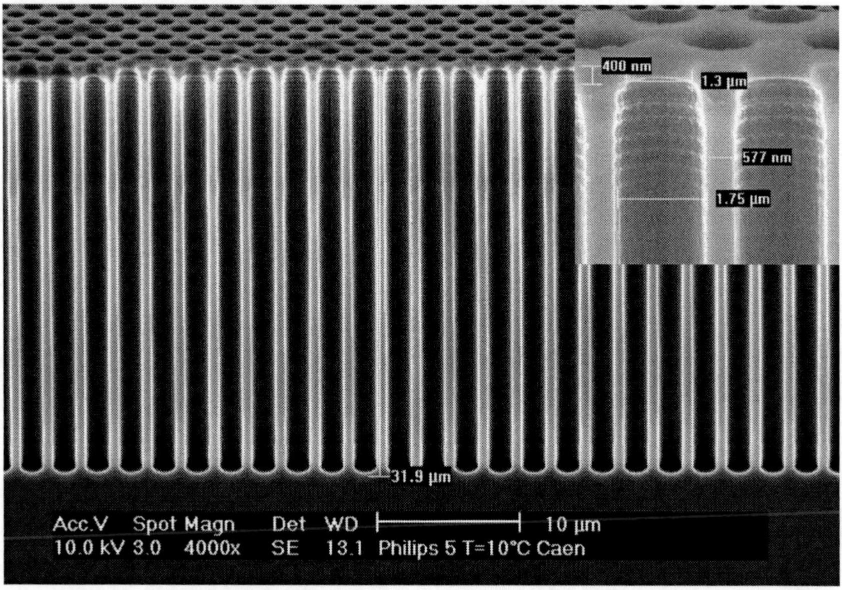

Figure 1. Array of micro-pores etched in silicon using DRIE process to realize the area enhancement. The aspect ratio in this typical example is ~13. The scallop structures on the sidewalls shown in the inset are due to repetitive Bosch process [F. Laermer and A. Schilp, "Method of anisotropically etching silicon,"U.S. Patent No. 5501893 March, 26, 1996].

Typically, the substrate patterning is done by Bosch process that involves ion etching together with SF_6/O_2 gas mixture, followed by a surface passivation step using C_4F_8 (without ions) to protect the sidewalls. Repetition of these steps yields high-aspect-ratio pores. The pores are normally positioned in a hexagonal lattice and separated far enough to avoid electrical crosstalk between the capacitors. The typical diameter of the pores is 1.5-2 μm with a spacing of 0.5-1 μm creating a pitch of 2-3 μm. Figure 2 shows the general schematics of the device fabrication. Step 1 is the substrate patterning with Bosch process, followed by the deposition of high-k layer and electrode layer. For conformal and pin-hole-free deposition of both layers, atomic layer deposition is specifically chosen, because other deposition techniques do not have the control in thickness and quality of material in large aspect-ratio structures. After the high-k and electrode deposition, both are etched and top and bottom contacts are made using sputter-deposited aluminium layers. In this way multiple metal-insulator-metal (MIM) devices can be fabricated on the same substrate and by connecting them in parallel, the total capacitance of the stack can be measured.

The micro-pore arrays dry-etched were with high aspect ratio up to 20. With these substrates, a capacitance density of 20 nF/mm^2 could be achieved, with an electrical breakdown voltage of 18 V for a ~20 nm Oxide-Nitride-Oxide (ONO) dielectric stack [16]. With ONO layer of 10nm-5nm-10nm thickness and pore-depth of 35 μm, a capacitance density of 30 nF/mm^2 has been achieved (see Figure 4(b)) [17]. Further on, 80 nF/mm^2 has been demonstrated with micro-pore capacitors using a thinner ON stack (though the breakdown voltage reduced to 8V) [18]. Although thinner dielectric films give larger capacitance densities, their breakdown voltages are lower, which limits their application scope. Moreover, such dielectrics are more vulnerable according to industrial standards, in terms of the durability of the devices. Therefore, multiple stacks of capacitors were contrived on top of each other and connecting in parallel, they yielded >200 nF/mm^2 with breakdown voltages of 11 V using multiple ONO stacks [19].

ONO layers were replaced by high-k layers deposited by atomic layer deposition to deposit thin dielectric layers together with electrode films in high-aspect-ratio pores. The deposition of good quality Al_2O_3 films by ALD is known in literature [20]. ALD synthesized Al_2O_3 in combination with TiN electrode layer has been studied in detail focusing on the high-aspect ratio micro-pore devices and a capacitance density of 180 nF/mm^2 has been attained using MIMIM stacks with 10 nm Al_2O_3 layers [20, 21, 22]. As mentioned earlier, the micro-pores were etched 20-30 μm deep into the n^{++} Si substrate, with 1.5 μm diameter and 3.5 μm pitch in footprint area varying from 120×120 μm^2 to 6000×6000 μm^2. After Al_2O_3 deposition, some wafers were annealed *in situ* in O_3 (for 300 s at 400°C) to improve the dielectric isolation and breakdown properties. The first layer was 5 nm SiO_2 over the Si substrate, on which 30 nm TiN was deposited by means of ALD. Two stacks of 10 nm

Al_2O_3/30nm TiN were deposited successively on this to form 2 stacks of metal-insulator-metal (MIM) structures, which formed a MIMIMIM device configuration. Figure 3 shows the TEM image of the step-coverage of these films over different regions of the micro-pore devices.

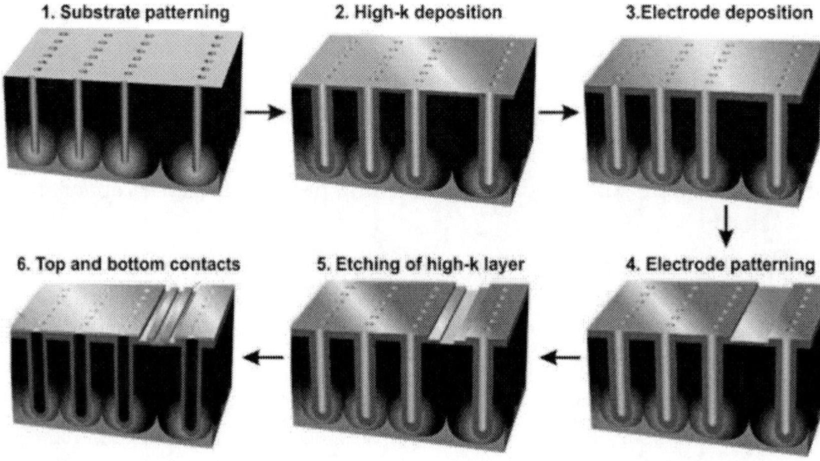

Figure 2. Schematic process flow of realizing the high-density capacitors.

With the optimized processing, the MIMIMIM stacks comprising one SiO_2 and two Al_2O_3 stacks could provide an ultra-high capacitance density of 450 nF/mm^2, which was the highest capacitance density in the world reported until 2008, and more than twice the value achieved using ONO stacks in the previous PICS program [23]. A capacitance density of 1 μF/mm^2 would be therefore plausible by replacing Al_2O_3 with a high-*k* material of *k*~20, by replicating the same process for the high-*k* layers. However, despite the challenges of depositing the high-*k* layers in high-aspect ratio micro-pores, there are several technical challenges to be solved. The deposition of Al_2O_3 in deep pores appears to be possible with appreciable step-coverage even at the bottom of the pores, but the area enhancement factor reduces with increasing the film thickness.

1.3. High-*k* Materials in Trenched Geometries

Despite the increased area, the parameter that can be tuned to achieve a large capacitance density is the dielectric constant *k* of the material, and a lot of research is ongoing in this field to discover the high-k dielectric materials with k values much larger than the conventional SiO_2 (k~3.9). Several material combinations have been emerged, starting from Si_3N_4 (k~5.6 -6.3) [24], Al_2O_3 (k~9) [25], ZrO_2 (k~22) [26], HfO_2 (k~25) [27], rare-earth metal oxides including the oxides of Lanthanide series (k~11-27), TiO_2 (k~80) and tertiary metal oxides such as $SrTiO_x$ with k>200 [28].

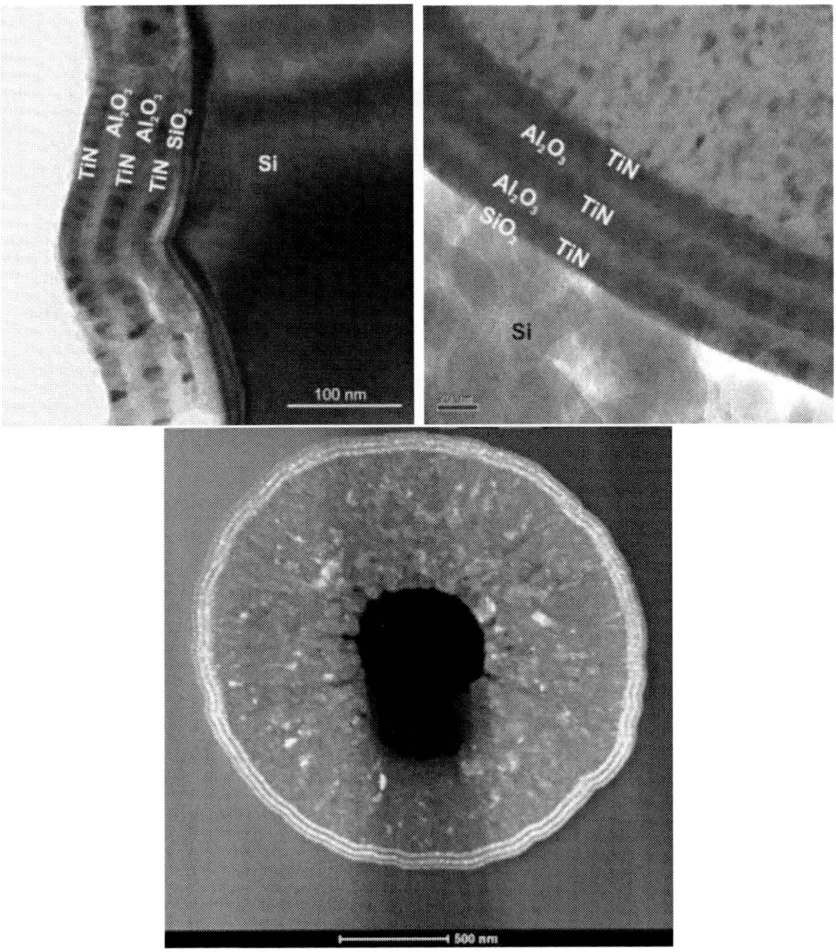

Figure 3. High resolution TEM of MIMIMIM stack deposited over micro-pore structures in silicon substrate. (Left): uniformity of Al_2O_3 and TiN layers deposited by atomic layer deposition (ALD) on the sidewalls and (middle) on the bottom surface of the pores. (Right): the high angle annular dark field (HAADF) image of the pore, showing the excellent coverage of the layers in the pore.

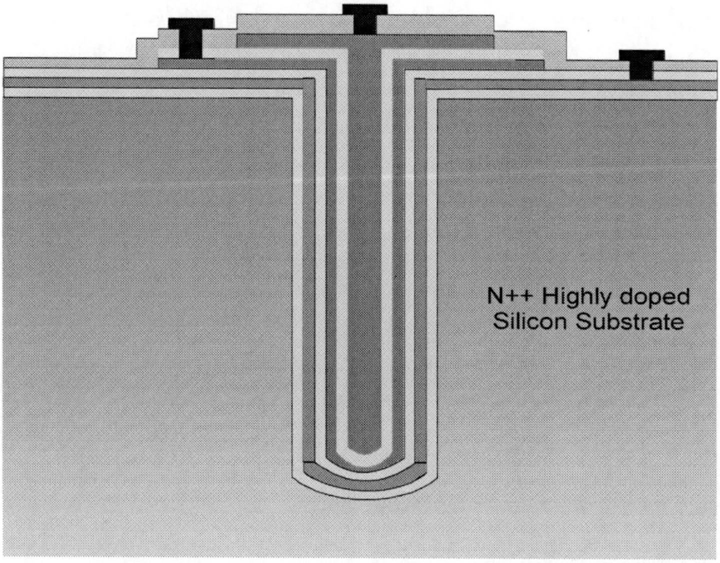

(a)

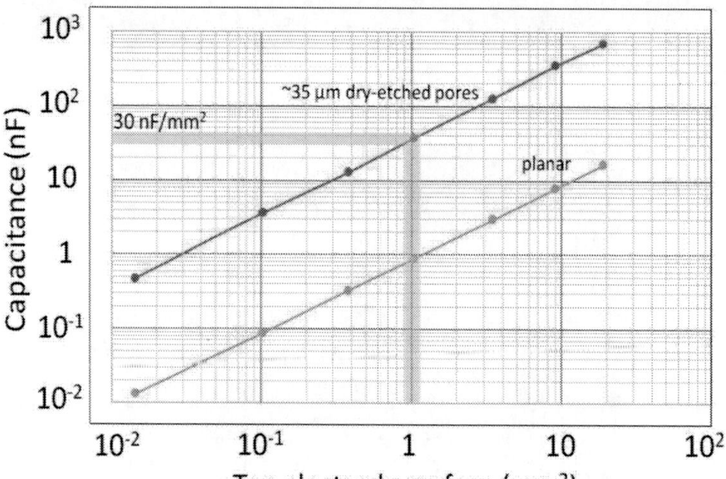

(b)

Figure 4. (left) The schematics of the connection to successive layers for capacitance measurements. (Right):capacitance with the foot-print area for MIM (with ONO layers) using pores of aspect ratio ~ 30. The capacitance enhancement of ~30 is clearly visible.

Subsequent to the success of MIMIMIM stacks with Al_2O_3 films, the deposition of $La_yZr_{1-y}O_x$ films in deep pores were performed to study the step-coverage and the electrical properties of MIS capacitors. $La_yZr_{1-y}O_x$ films exhibits k-values as high as 36 [29]. Pre-patterned silicon substrates with 10 μm deep pores were subjected to thermal oxidation to create a 5nm thick SiO_2 layer, on which $La_yZr_{1-y}O_x$ and TiN films as electrodes were deposited by ALD. HRTEM images of top, sidewalls and bottom of the pores shown in Figure 5 illustrates very conformal deposition of the layers over the pore surface.

The deposited layers have excellent step-coverage and TiN is polycrystalline throughout the surface of the pores, which indicates that the layers are of the same quality both on the surface and in deep pores. Figure 6 shows the capacitance enhancement due to the trenched substrates.

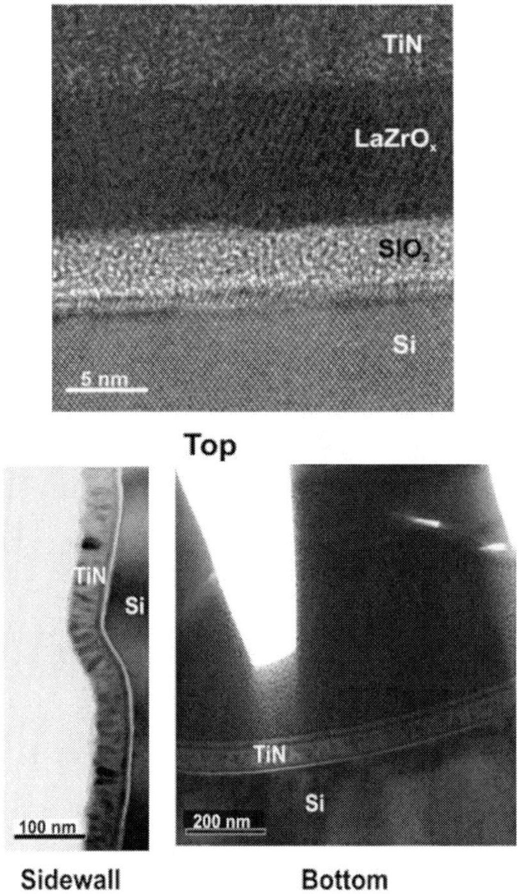

Figure 5. HRTEM images of the $LaZrO_x$ film deposition on top, sidewall of the pore and bottom of the pore.

1.4. RELIABILITY ASPECTS OF 3D CAPACITORS

Trenched geometry gives option for area enhancement to a large extend. However, the breakdown voltage is consistently lower for the micro-pore capacitors compared to the planar capacitors. This observation suggests that micro-pore capacitors have a shorter lifetime at the same bias voltage than the planar devices. Figure 7 shows a set of Weibull distributions of capacitor breakdown voltages. Since the slopes of the Weibull distribution of the micro-pore devices are comparable to those of the planar devices, it can be understood that the earlier breakdown of the micro-pore devices are not due to extrinsic failure resulting from processing non-uniformities of the devices, but are intrinsic failures related to the dielectric breakdown of the films.

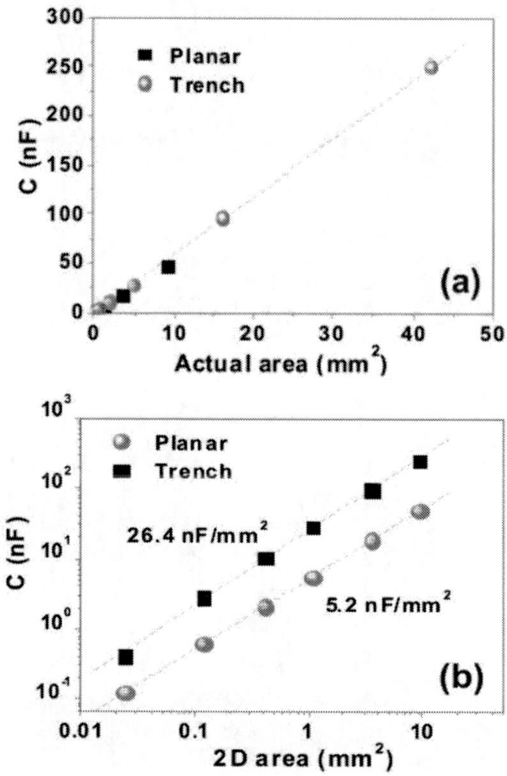

Figure 6. (a) Capacitance of $La_yZr_{1-y}O_x$ capacitors on planar and trenched silicon substrate scaling with the actual area of the devices; (b) capacitance-scaling with footprint area of the devices.

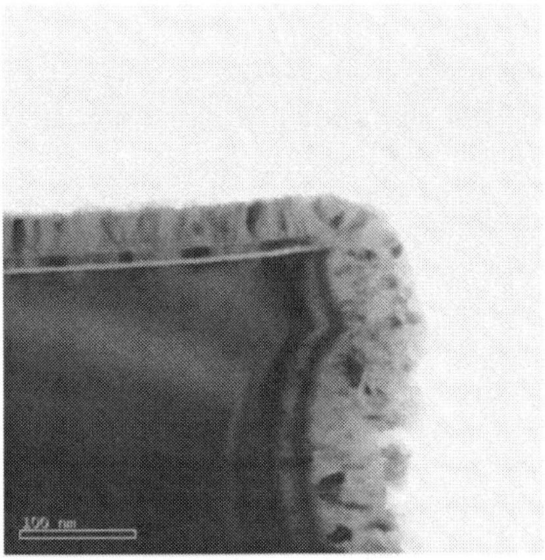

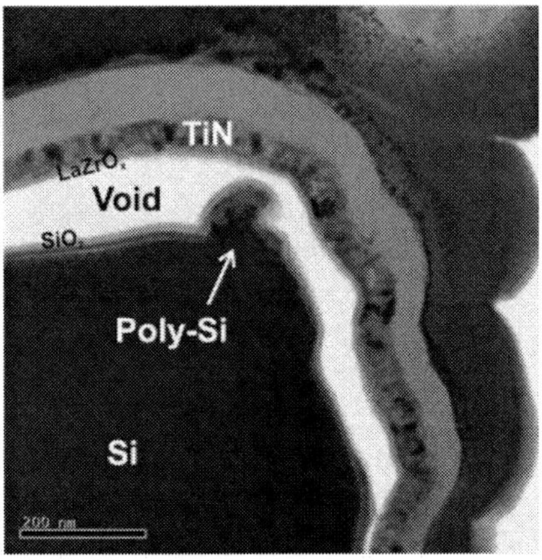

Figure 7. High resolution TEM images of the corners of micro-pore devices before and after dielectric breakdown, showing the rupture of the layers at the corner. The $La_yZr_{1-y}O_x$ layer together with the TiN film is completely delaminated at the corner, leaving out a void (white region in the picture on right), while the SiO_x reduced to 0.6 nm. Monocrystalline Si substrate turned to polycrystalline Si at the corner where the breakdown occurred.

For the same applied voltage, the sharp edges of the pores generate enhanced electric field that give rise to larger leakage currents and earlier onset of Fowler-Nordheim (FN) tunnelling [30] suggesting that the electric field causing the dielectric breakdown is larger for micro-pore devices than for the planar devices. When the layer thickness is smaller than the radius of curvature of the corner (or any other sharp edges), there is a uniform electric field inside the high-*k* layers. But when the thickness of the layer is larger than the radius of curvature of the corners, the electric flux density is larger at the sharp edges (as per Gauss' theorem). This field enhancement at the sharp edges results in earlier FN tunnelling at same applied voltages than the planar devices, causing earlier device breakdown.

When the dielectric breakdown happens, a mark due to the rupture of the layers appears on the top TiN electrode. A dedicated high resolution TEM (HRTEM) experiment was performed to discover the origin of the breakdown spot in micro-pore devices. These analyses show that the breakdown is indeed at the corner of the micro-pores, where the $La_yZr_{1-y}O_x$ film together with TiN layer appears to be delaminated from the Si surface, while the SiO_2 layer reduced from 4.2 nm to 0.6 nm on Si substrate (as seen in the right-side image in Figure 7), creating a void between the dielectric layers. The TEM analyses demonstrate that rupture of the film occurs all across the films and even in the silicon substrate. Technologically challenging feature is that the breakdown always happens at the pore-edges. Increasing the radius of curvature of the pore-corners by rounding the edges with an additional etching prior to the deposition of the high-*k* films would increase the breakdown voltage to similar to that of the planar devices.

1.5. FUTURE OF THE 3D CAPACITOR TECHNOLOGY

Decoupling capacitors are vital components that minimise cross-talk in radio frequency (RF) communications products, *e.g.* mobile phones, but they contribute substantially to manufacturing costs. 3D capacitor technology offers an elegant solution to integrate decoupling capacitors into integrated circuits and with the advancement in flip-chip technology, more and more components can be crammed into the chips, making the 3D integration of passive elements possible. This is a giant leap in system-level integration technology.

REFERENCES

[1] G. Moore, *Electronics* 38, 114 (1965).
[2] G. E. Moore, *Tech. Dig. IEEE Int. Electron Device Meeting*, page11, (1975).
[3] *International Technology Roadmap for Semiconductors (ITRS),* 2011 Edition.
[4] M. Hirose et al., *Semicond. Sci. Technol.* 15, 485 (2000).
[5] D. A. Muller et al., *Nature* 399, 758 (1999).
[6] R. Landauer, *IBM J. Res. Dev.* 5, 183 (1961).
[7] J. Izydorczyk, *IEEE Trans. Very Large Scale Intergr. (VLSI) Syst.* 18, 161 (2010).
[8] M. de Samber, L. Tegelaers, Philips. *J. Res.* 51, 389 (1998).
[9] F. Murray et al., *Advanced Electronic Packaging Symposium*, page 3 (2007).
[10] D. Chevrie, et al., *IMAPS Conference,* Bruges, Belgium, 2005 (June 12–15.
[11] F. Roozeboom et al. *Thin Solid Films* 504, 391 (2006).

[12] J. D. Jackson, *Classical Electrodynamics*, 3rd ed. Wiley, NewYork, (1999).
[13] M. Lenzlinger, E. H. Snow, *J. Appl. Phys*. 40, 278 (1969), G. Pananakakis, et al., *J. Appl. Phys*. 78, 2635 (1995).
[14] H. C. Lin, P. D. Ye, G. D. Wilk, *Solid-State Electron*. 50, 1012 (2006).
[15] V. Lehmann, U. Grüning, *Thin Solid Films* 297, 13 (1997).
[16] F. Roozeboom et al., *Int. J. Microcircuits and Electronic Packaging*, 24, 182 (2001).
[17] J. H. Klootwijk, K. B. Jinesh, F. Roozeboom, *Microelectronic Engineering,* 88, 1507 (2011).
[18] F. Murray et al., *Mater. Res. Soc. Symp*. 969, 27 (2007).
[19] F. Roozeboom et al., *ECS Transactions* 173, (2007).
[20] J. L. van Hemmen et al., *J. Electrochem. Soc*. 154, G165 (2007).
[21] D. J. Hoogeland et al., *ECS Transactions* 25, 389 (2009).
[22] S. B. S. Heil et al., *J. Vac. Sci. Technol. A* 25, 1357 (2007).
[23] J. H. Klootwijk et al., *IEEE Electron Device Lett*. 29, 740 (2008).
[24] S. A. Awan et al., *Thin Solid Films* 423, 267 (2003).
[25] M. D. Groner et al., *Thin Solid Films* 431,186 (2002).
[26] D. H. Triyoso et al., *Appl. Phys. Lett*. 88, 222901 (2006).
[27] G. D. Wilk, R. M. Wallace, J. M. Anthony, *J. Appl. Phys*., 87, 484 (2000).
[28] J. Robertson, *Eur. Phys. J. Appl. Phys*. 28, 265 (2004).
[29] K. B. Jinesh et al, *Appl. Phys. Lett.* 93, 062903 (2008).
[30] P. H. Cutler, J. He, N. M. Miskovsky, T. E. Sullivan, B. Weiss, *J. Vac. Sci. Technol*. B 11, 387 (1993).

In: Capacitors
Editor: Kristofer N. Muller

ISBN: 978-1-62417-585-5

Chapter 5

IMPACT OF THE CAPACITORS ENERGIZATION OVER THE POWER QUALITY OF THE MV DISTRIBUTION GRIDS

Jorge Blanes*

Department of Electrical and Automatic Engineering, University of León
Campus de Vegazana s/n, León, Spain

ABSTRACT

The connection of capacitor banks at the MV distribution network busbar feeder is a technique widely used by the Spanish utilities. It decreases the reactive power transit (diminishing the losses) and improves the voltage profile along the distribution lines. However, this technique has a negative influence over the Quality of Supply because the capacitor's energization generates an oscillatory transient surge that propagates along the lines little damped which can cause problems in the MV elements and in the LV equipments. To analyse the generation and propagation of these surges, several power monitors were installed at different sites of the MV distribution network. A Fourier analyse of the waveforms recorded will show the frequencies at different sites revealing the 'natural' frequencies of the users (including the MV/LV transformer and their LV loads).

* Tel: +34 987 29 1990. Fax: +34 987 29 1770. E-mail: jorge.blanes@unileon.es.

Finally, the techniques to reduce the amplitude and the negative effects of these surges will be shortly presented.

Keywords: Capacitors energization, power quality, transient surges

1. Introduction

In recent years, the installation of capacitor banks at the busbar feeder of the Spanish 20 kV distribution network has become the norm [1, 2]. This technique contributes to an efficient management of the system since it decreases the reactive power transit (power factor correction) which means lower losses and improves the voltage profile along the distribution lines. To optimize the benefits an adequate strategy must be retained resulting in a compromise between the loss reduction and the increasing costs of the compensation (capacitor banks and the associated equipment). The drawback of this technique is related to the capacitor banks energization which generates an oscillatory transient surge that can spread along the lines before it eventually damps out. To analyse the generation and propagation of these surges, several power monitors were installed at different sites of the MV distribution network.

2. Network Description and Power Monitors Specifications

Figure 1 shows the three-phase network diagram. This representation has been used instead of the conventional single-phase diagram because it shows better the connection of both the loads and the capacitor bank. The most representative characteristics of the network are:

- Source frequency: 50 Hz
- Strength of the HV System at the 66 kV busbar: 900 MVA
- HV/MV transformer: Three phase, 20 MVA power rate, two windings 66±10% /20 kV, Ynd11 connection
- Capacitor bank: 23.68 μF, wye connection (with no neutral grounding).
- Line A: 3 km overhead line

- Rest of lines: Represents the rest of the lines departing for the 20 kV busbar feeder
- MV/LV transformer: Three phase, 800kVA power rate, two windings 20.000/400 V, Dyn11 connection
- Loads (power consumptions):
  - Total 20 kV feeder loads:
    - Active Power: 13.3 MW
    - Reactive Power: 4.37 MVAr
  - Line A total loads:
    - Active Power: 5.06 MW
    - Reactive Power: 1.65 MVAr
  - User (object of study) loads
    - Active Power: 3.26 MW
    - Reactive Power: 1.07 MVAr

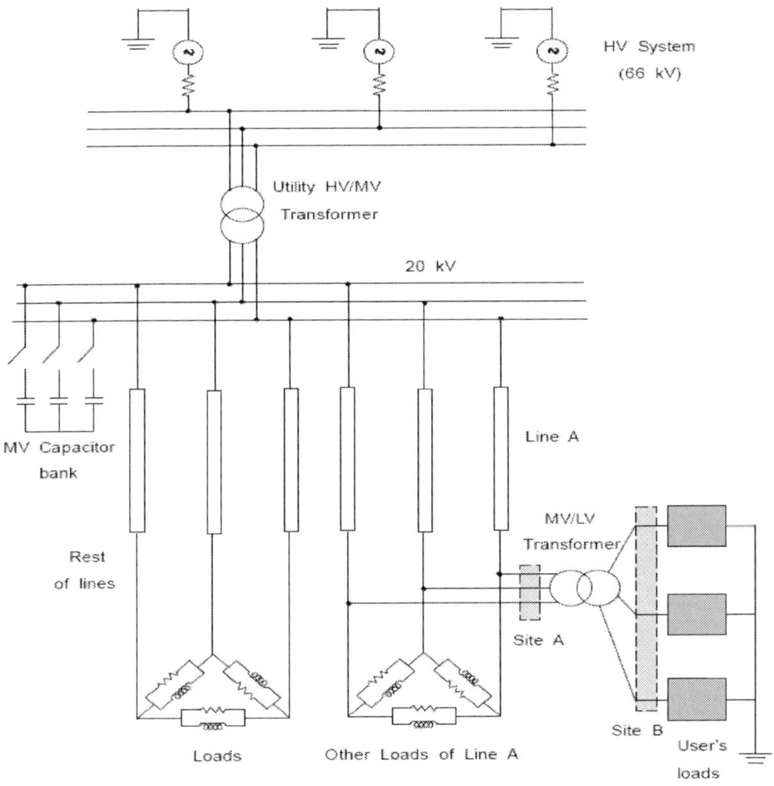

Figure 1. Three-phase network diagram (the LV loads are represented as black boxes).

Two power monitors were connected at sites A (MV measuring point) and B (LV measuring point). Their main specifications are:

- Voltage measurement:
  - Accuracy: ±0.02 full scale ±0.2 of reading
  - Sample rating: 256 samples per cycle.
- Current measurement:
  - Accuracy: ±0.04 full scale ±0.2 of reading
  - Sample rating: 128 samples per cycle.

3. Recorded Waveforms

Figure 2 shows the voltage and current waveforms recorded in the LV side of the network (site B of figure 1) during a transient event produced by the MV capacitor energization. A closer inspection of the waveforms reveals an abrupt drop of the voltage at the instant of switching. This is related to the charge of the capacitor bank which behaves initially as a short circuit.

4. Analysis in Frequency Domain

The objective of this analysis is to determine the frequencies in the LV voltage and current waveforms due to the capacitor bank energizing. A preliminary signal processing [3, 4, 5] is carried out to obtain the maximum frequency range for which the measurements are acceptable [6, 7, 8, 9]:

1. For each phase the auto power spectra functions of the voltage and current signals i.e. $\Phi_{vv,n}(w)$ and $\Phi_{ii,n}(w)$ (where $n = a, b, c$) are processed with the aid of the Fourier Transform method of the autocorrelation functions. To achieve the former the *Welch* estimator is used, working with six sequences of 256 samples.
2. For each phase, the cross power spectra functions between voltage and current $\Phi_{iv,n}(w)$ ($n = a, b, c$) are processed and then the coherence function is obtained from the following expression:

$$\Gamma^2_{iv,n}(w) = \frac{|\Phi_{iv}(w)|^2}{\Phi_{vv}(w) \cdot \Phi_{ii}(w)} \quad (n = a, b, c) \tag{1}$$

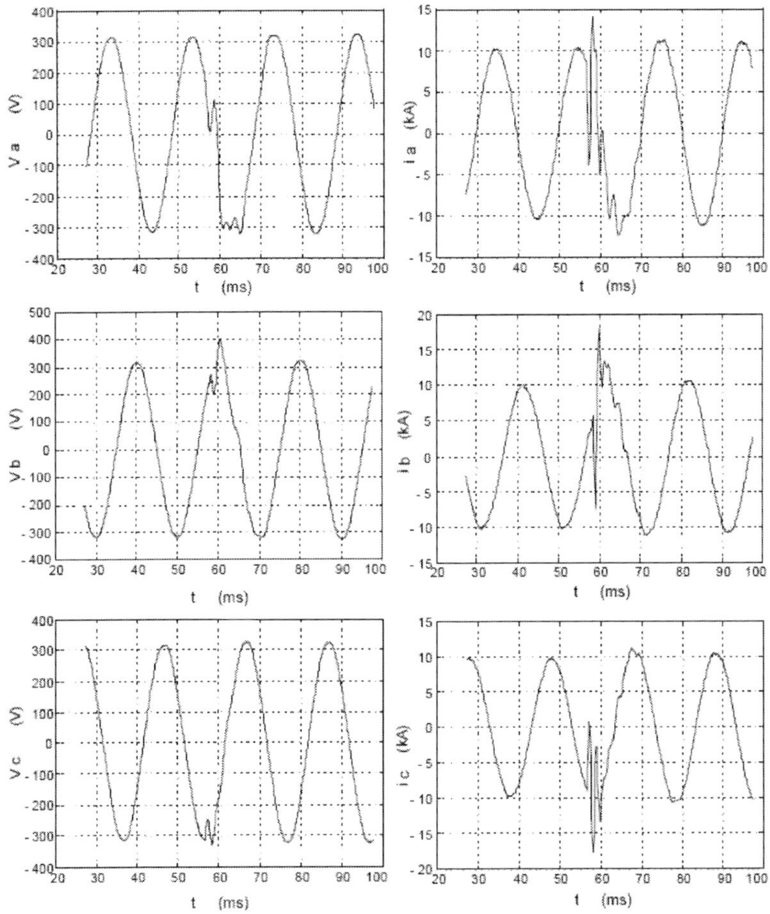

Figure 2. Waveforms recorded in LV.

Figure 3 shows the modules of these functions.

The analysis of the last figure shows that the modules of the coherence functions decrease abruptly for frequencies greater than 850 Hz. So this value represents the maximum frequency for which there is a relationship (cause/effect) between the voltage and current recorded waveforms.

The next step is to obtain the FFT of the LV waveforms for the period before the energization of the capacitor bank. This provides knowledge of the existing frequencies in steady state which are not attributable to the transient. Then the FFT of the global waveforms (including the capacitor energizing period) are carried out. Figure 4 shows the modules of the resulting averaged values (for the three phases).

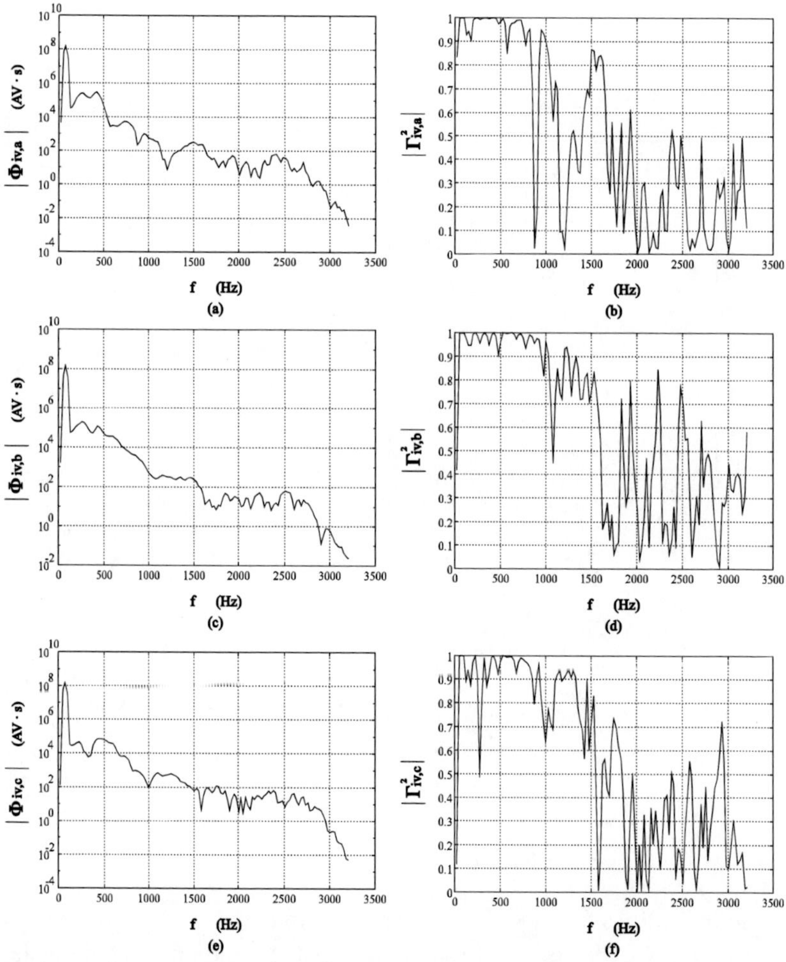

Figure 3. Modules of the crossed spectral density and coherence functions of the LV signals. (a) Phase a crossed spectral density function. (b) Phase a coherence function. (c) Phase b crossed spectral density function. (d) Phase b coherence function. (e) Phase c crossed spectral density function. (f) Phase c coherence function.

To establish the origins of these frequencies it is necessary to compare them with those of the waveforms recorded on the MV side (site A in figure 1) which are shown in figure 5. The analysis of this figure shows that, ignoring the frequencies in the steady condition, the dominant frequency spectrum for both voltage and current waveforms is around 230 Hz whereas at 400 Hz the voltage spectrum disappears completely. Based on these findings the following hypotheses concerning the frequencies are outlined:

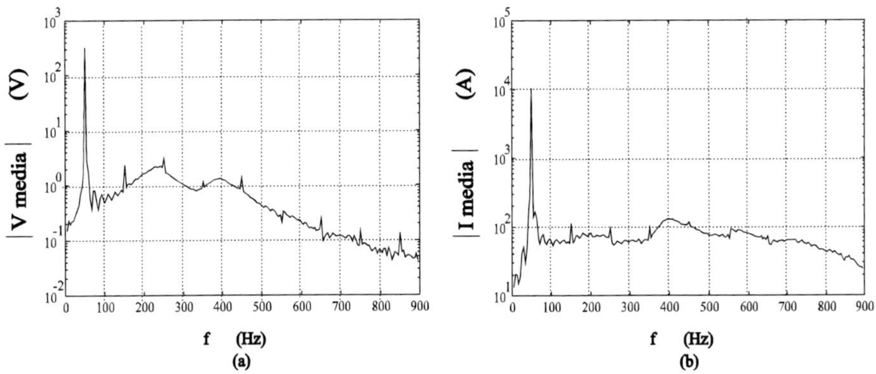

Figure 4. Modules of the FFT of the LV waveforms: Averaged values for the modules of the three phases LV voltages ($|V_{media}|$) and LV currents ($|I_{media}|$) of the waveforms recorded at site B.

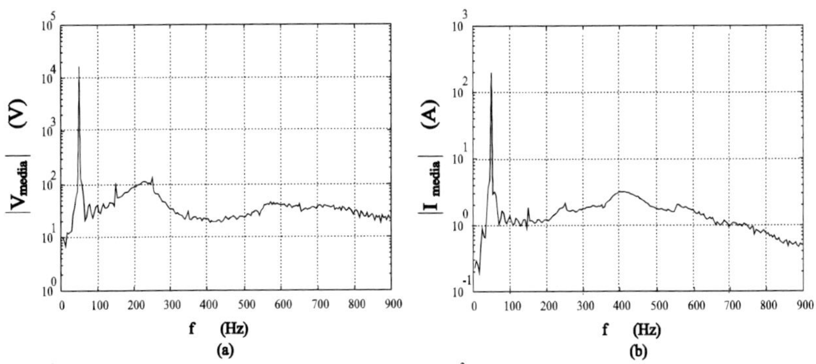

Figure 5. Modules of the FFT of the MV waveforms: Averaged values for the modules of the three phases MV voltages ($|V_{media}|$) and MV currents ($|I_{media}|$) of the waveforms recorded at site A.

- 230 Hz is the forced response of the LV load to the transient phenomena from the MV side. This value is due to the actual network configuration and the capacitor bank.
- 400 Hz is due to the natural response of the users (including the MV/LV transformer and their LV loads).

It is essential to know precisely the last value (natural frequency) as it may appear a resonance phenomenon that amplifies the problem. The following section describes an analysis in time domain that will allow us to characterize this frequency with more precision.

5. Analysis in Time Domain

In this analysis the decomposition of the voltage and current waveforms is carried out using Prony's method. It is known that when this technique is applied to signals with non-damped components, it leads to incorrect decompositions. To avoid these errors a digital filtering of the signals is carried out for elimination of the permanent component of 50 Hz. The digital filter used is a high pass *Butterworth* filter that eliminates frequencies lower than 50 Hz.

Table 1.

	f (Hz)	α (s^{-1})	A (V)	φ (rad)		f (Hz)	α (s^{-1})	A (A)	φ (rad)	
	142.79	-3416	3.63	-0.87		178.23	-223.32	2056.97	-2.09	
v_a	196.91	-22.22	2.62	-2.49		403.74	-211.71	4013.23	2.12	i_a
	232.78	-146.19	57.03	-3.08		801.73	-291.18	2197.98	0.92	
	403.57	-203.05	46.15	0. 31		168.64	-5.85	98.34	-0.48	
	153.74	18.70	0.55	-2.80		173.79	-156.47	2664.33	2.06	i_b
v_b	235.86	-134.82	42.60	0.05		427.38	-281.28	4474.65	-1.40	
	422.89	-161.04	27.15	2.53		691.68	-53.36	906.67	1.21	
	191.10	-110.95	12.73	0.41		92.33	-92.16	532.12	0.65	
v_c	271.56	-42.51	2.98	-2.39		372.05	-33.61	438.21	0.16	i_c
	334.25	-11.95	1.23	2.04		379.64	-30.98	406.59	1.78	
	438.49	-324.85	31.28	-1.58		585.46	-363.19	5589.77	-1.55	

Prony's method allows the decomposition of a general signal such as $x(t)$ which has N elements and it is expressed as:

$$x(t) = \sum_{i=1}^{N} A_i \, e^{-\alpha_i t} \cos(2\pi f_i + \varphi_i) \tag{2}$$

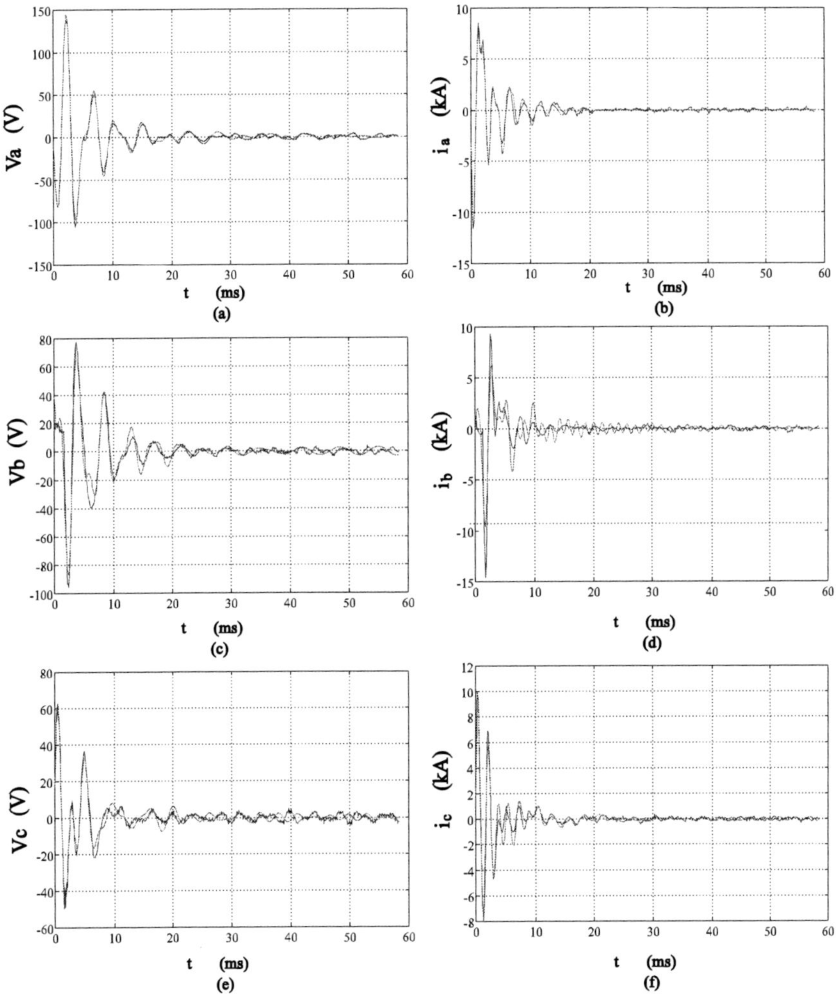

Figure 6. Result of Prony's decomposition method: Original filtered signals (blue) and signals resulting from the addition of Prony's components (red).

Each element of the waveform is defined by four parameters: A_i (amplitude), α_i (damping coefficient), f_i (frequency) and φ_i (phase). Table 1 shows these parameters for each phase of the LV filtered signals. Figure 6 compares the original signals with those obtained using Prony's decomposition.

To determine the predominant frequency around 400 Hz, the pondered frequencies of voltages and currents are obtained using amplitudes as a ponder factor. These are:

$$f_v = \frac{\sum\limits_{n=a,b,c} f_{v,n} \cdot A_{v,n}}{\sum\limits_{n=a,b,c} A_{v,n}} = 419.03\,Hz \qquad f_i = \frac{\sum\limits_{n=a,b,c} f_{i,n} \cdot A_{i,n}}{\sum\limits_{n=a,b,c} A_{i,n}} = 416.20\,Hz \tag{3}$$

Finally, the average of these two values gives 417.61 Hz.

6. Techniques to Minimize the Effects of the Surges

Some researchers [10, 11] found that the peak value of this voltage surge can reach twice the rated voltage causing problems both in the MV side (lines and transformers switching off) and in the LV side (leading to electrical protections malfunction, electrical drives and other electronic equipment failures,...). For this reason diverse techniques have arisen to minimize the effects of these surges. They can be classified in two big groups:

- Techniques to reduce the amplitude of the switching surge. Their purpose is to minimize the source of the problem, acting, therefore, over the element generating the transient (utility capacitor energization).
- Techniques to reduce the effects of the surge which are applied in LV.

6.1. Techniques to Reduce the Amplitude of the Surges

Oscillatory transient surge phenomena have been widely studied by many researchers as well as the techniques used to reduce the oscillatory transient surge. The main techniques are:

Pre-Insertion Resistors

This technique reduces considerably the magnitude of the switching surges but presents two main problems when taking it to the practice:

- The optimum size resistors are frequently of non standard values, being necessary an special production.
- The resistors are only present at the switching instant since, if they were permanently connected, they would produce important losses. Is then necessary an additional switch of by-pass that rises the cost of the device.

Pre-Insertion Inductors

This is an attractive solution in terms of cost and reliability so, it is displacing the preceding one. Anyway, it is a technique not always effective [12]. The effectiveness depends notably on the network configuration since the new inductance decreases considerably the transient frequency.

Permanent Insertion Inductors

An inductor of some dozens of mH (filtering inductance) in series with the capacitor bank may reduce the amplitude of the surges until non harmful values. This solution can not be adopted if the capacitor has not been designed to support the permanent higher voltage values caused by the inductor [13].

However, a coil of only some tenth of mH (choke) in series with the capacitor can limit the transient current through them but it can not reduce the amplitude of the surge significantly.

The reference [13] presents an interesting study about the possibility of using inductors of an intermediate size between the two previous that can considerably reduce the amplitude of the surges without causing a permanent increase of the voltage. This study was carried out using a numerical simulation program and the first results have not been very encouraging since, to reduce in 50% the amplitude of the surge it was necessary an inductor that originated a permanent increase of the voltage of about 2.5%.

Synchronized Switching

This technique consists in closing each of the poles of the capacitor at a zero of the voltage across the switching device [14]. This solution was recognized as preferential topic in the session of 1990 of the CIGRE and gave place to the creation of a Committee charged of its study: the Study Committee 13 (SC-13) [15].

The main practical problems of this technique are the appearance of the arch which causes the effective closing of the poles before its physical contact and the probability of restriking. If restriking is presented (not with the modern SF_6 switching devices) the battery can be loaded to a certain voltage level. Therefore, the closing of the poles should not be at a zero of the voltage across the switching device but when this value is equal to the capacitor voltage.

6.2. Techniques to Reduce the Effects of the Surge

To aggravate the problem, in some circumstances, magnifying phenomena can occur at the LV users' installations. There are two "magnifying" mechanisms: The first is due to the reflection at the end of the MV line. The second is a consequence of a resonant circuit formed between the MV/LV users' transformers and their own LV capacitor banks [16, 17], generating voltage surges, in some cases, as high as five times the rated voltage. These capacitors can be the LV power factor correction capacitors or, even, the, sometimes, large capacitors used to maintain a constant voltage in the asynchronous speed drives that use voltage source inverters. In this case an optimal solution is the insertion of inductors in series with the LV capacitors, forming an harmonic filter capable of detuning the resonance of this circuit series of the main frequency of the transient surge coming from MV. Anyway, the low voltage metal oxide varistors (MOV's) should always be used to protect sensitive equipment. If the preceding phenomenon of magnification in LV of the oscillatory switching surge is presented, these arresters should be designed to deal with energies of about 1000 Joules [18].

Conclusion

If the main frequency of the oscillatory switching surge is close to the natural frequency of a user, which is obviously time dependent, important damages due to a resonance effect may be attended. To avoid this negative situation, it would be useful an intermediate solution of the mitigations techniques exposed in this paper consisting in an intelligent control capable of a permanent survey of the network/users configuration. This will allow the synchronized switching of the capacitor with inductors in series which values are in each situation the optimal not only reduce the amplitude of the surges but to separate its main frequency from the 'dangerous' values.

REFERENCES

[1] L. de la Cuadra, F. Cavallé, A. Quijano, J. Blanes, V. Fuster, C. Cañas, 'Short time interruptions project: Their causes and effects on MV networks', *Second International Conference on Power Quality: End-Use Applications and Perspectives (PQA'92)*, Atlanta, USA, September 1992, pp. B-21:1-8.

[2] J. Blanes, 'Influence of the Capacitors Energization over the Power Quality of the Distribution System', *33rd Universities Power Eng. Conf. (UPEC'98)*, Edinburgh, 8-10 September 1998, 631-634.

[3] M. Kunt, 'Traitement numérique des signaux', 3e édition, *Éditions Presses Polytechniques Romandes*, 1981.

[4] A. V. Oppenheim, R. W. Schafer, 'Discrete-Time Signal Processing', *Prentice Hall International, Inc.*, Englewood Cliffs, New Jersey.

[5] L. P. Rabiner, B. Gold, 'Theory and Application of Digital Signal Processing', *Prentice Hall International, Inc.*, Englewood Cliffs, New Jersey.

[6] A. S. Morched, P. Kundur, 'Identification and Modelling of Load Characteristics at High Frequencies', *IEEE Trans. on Power Systems*, Vol. PWRS-2, No. 1, (1987) 153-160.

[7] A. Robert, F. Schekler, 'Guide for Assessing the Network Harmonic Impedance (1st draft)', *Working Group CC02, CIGRE 36.05, CIRED 2*, paper No. 9244, 1992.

[8] J. S. Bendat, A. Piersol, 'Engineering Applications of Correlation and Spectral Analysis', *John Wiley and Sons, Inc.*, New York, 1980.

[9] J. Blanes, H. Nouri, 'Synthesis of LV load Models for MV Networks Fast Transient Studies', *Electr. Power Syst. Res.*, 2011, doi:10.1016/j.epsr.2011.10.007.

[10] T. Henriksen, A. P. Brede, 'Overvoltage Disturbances Caused by the Energization of a Capacitor Bank', *Working Group CC02, CIGRE 36.05, CIRED 2*, paper No. 9194, 1991.

[11] M. F. McGranaghan, D. Mueller, W. H. Jones, S. P. Conrad, B. W. Buckner, 'Evaluating Customer-Side Transients Caused by Utility Capacitor Switching', *First Eur. Conf. on Power Systems Transients (EPST'93)*, Lisbon, 17-18 June 1993, 259-263.

[12] B. Bhargava et al., 'Effectiveness of the Pre-Insertion Inductors for Mitigating Remote Overvoltages Due to Shunt Capacitor Energization', *IEEE Transactions on Power Delivery*, Vol. 8, No. 3, pp. 1226-1238, July 1993.

[13] Laborelec, 'Influence d'une self serie sur le transitoire d'enclenchement d'une baterie de condensateurs MT', *Working Group CC02, CIGRE 36.05, CIRED 2*, paper No. 9173, 1991.

[14] R. Söderström, 'Synchronized Switching of MV-Circuit Breaker for MV-Capacitor Banks', *Third Int. Conf. on Power Quality: End-Use Applications and Perspectives (PQA'94)*, Session E-2: Mitigation Techniques, paper No. E-2.02, Amsterdam, October 24-27, 1994.

[15] CIGRE Study Committee 13. Task Force 13.00.1, 'Report on Controlled Switching', *Working Group CC02, CIGRE 36.05, CIRED 2*, paper No. 9169, 1991.

[16] M. F. McGranaghan, R. M. Zavadil, G. Hensley, T. Singh, M. Samotyj, 'Impact of Utility Switched Capacitors on Customer Systems. Part I: Magnification at Low Voltage Capacitors', *IEEE Trans. on Power Delivery*, Vol. 7, No. 2, (1992) 862-868.

[17] A. J. Schultz et al., 'Magnification of Switching Surges', *IEEE Transactions on Power Apparatus and Systems*, Vol. 77, pp. 1418-1426, February 1959.

[18] R. Alvinsson, C. E. Sölver, 'Switching to Lower Transients', *Transmission and Distribution, Working Group CC02, CIGRE 36.05, CIRED 2*, paper No. 9226, 1991.

INDEX

D

N

O

P

Q

R

S